California Dreaming

SUNY series in Israeli Studies

Russell Stone, editor

CALIFORNIA DREAMING

Ideology, Society, and Technology in the Citrus Industry of Palestine, 1890–1939

NAHUM KARLINSKY

Translated from the Hebrew by Naftali Greenwood

STATE UNIVERSITY OF NEW YORK PRESS

The publication of this book was made possible with the support of the Ben-Gurion University of the Negev.

Cover art: View of Emek Hefer, 1936. Photograph used with permission of the Israeli Government Press Office. Photographer: Zoltan Kluger.

Published by
State University of New York Press
Albany

For information, contact
State University of New York Press
Albany, NY   www.sunypress.edu

Production, Laurie D. Searl
Marketing, Susan Petrie

Library of Congress Cataloging-in-Publication Data

Karlinsky, Nahum.
[Perihat he-hadar. English]
California dreaming : ideology, society, and technology in the citrus industry of Palestine, 1890–1939 / Nahum Karlinsky.
p. cm. — (SUNY series in Israeli studies)
Includes bibliographical references and index.
Based on author's doctoral thesis, Hebrew University, 1995.
ISBN 0-7914-6527-6 (Hardcover : alk. paper)
ISBN 0-7914-6528-8 (Paperback : alk. paper)
1. Citrus fruit industry—Palestine—History. 2. Fruit trade—Palestine—History. 3. Entrepreneurship—Palestine—History. I. Title. II. Series.

HD9259.C53P34 2005
338.1'743'09569409041—dc22

2004022564

10 9 8 7 6 5 4 3 2 1

For my wife, Gannit,

and for our children—

Lee-Or, Roi, and Amir

with all my love.

Contents

List of Maps and Figures ix

List of Tables xi

Preface xiii

One The Tantalizing Aroma of Citrus Blossoms 1

Part I. Ideological Platform

Two Degania or Petah Tikva?: Private Enterprise in the Worldview of Jewish Citrus Growers in Palestine and Their Opponents 21

Part II. Social and Geographical Platform

Three Spatial Distribution and Social and Entrepreneurial Profile 49

Part III. Technological Innovations

Four From Jaffa to Petah Tikva: Technological Development in Citriculture during the Ottoman Period 87

Five Technological Innovations during the Mandate Era 111

Six Technological Innovations in Arab Citriculture 155

Part IV. Growing Pains

Seven Pursuit of Profit 167

Eight The British Mandate Government's Policy toward the Citrus Industry 181

Nine Attempts to Establish a Cartel 187

Ten Conclusion: Jewish Citriculture as a Private-Enterprise Industry 217

Notes 221

Bibliography 247

Index 267

Maps and Figures

MAPS

1.1 Palestine in 1936 17

3.1 Spatial distribution of Jewish citriculture, 1914 55

3.2 Spatial distribution of Jewish and Arab citricultures, 1935–1936 82

FIGURES

1.1 Nahum Gutman, *Orange Groves in Jaffa*, 1928 (photograph courtesy of the Nahum Gutman Museum) 3

1.2 "Area of Citrus Plantations and Export Figures, 1922–1938," Gurevich and Gerz, *Jewish Agricultural Settlement in Palestine* 14

2.1 Moshe Smilansky at fifty-three (photograph courtesy of Ms. Cila Agam) 25

2.2 David Ben-Gurion speaks before Jewish workers in Jerusalem, 1924 (photograph courtesy of the Lavon Institute for Labor Research) 38

3.1 Tovia Miller and Moshe Smilansky in front of Miller's house in Rehovot (photograph courtesy of Ms. Yael Miller, photographer: Mr. Schleisner, Rehovot) 61

3.2 View of Emek Hefer with an orange grove planted by Yakhin, 1936 (photograph courtesy of the Israeli Government Press Office, photographer: Zoltan Kluger) 73

4.1 Traditional water pumping system (photograph courtesy of Ms. Nava Chen) 93

5.1 Jewish demonstrators outside a private-owned orange grove in Kfar Saba, 1934 (photograph courtesy of the Lavon Institute for Labor Research) 113

5.2 Decoville in large Jewish Grove, late 1930s (photograph courtesy of the Israeli Government Press Office, photographer: Zoltan Kluger) 121

5.3 Manual irrigation (Jewish grove) (photograph courtesy of the Israeli Government Press Office, photographer: Zoltan Kluger) 130

5.4 The "California irrigation system" in Yakhin grove, 1936 (photograph courtesy of the Central Zionist Archives) 132

5.5 Picking at "modernized" Jewish grove, 1930s (photograph courtesy of the Central Zionist Archives) 137

5.6 Traditional packing at an Arab-owned grove (photograph by Elia Kahvedjian, courtesy of Elia Photo Service) 139

6.1 Citrus Marketing Board, 1940. G. Sandford (Financial Secretary), Isaac Rokach, G. Walsh (Economic Advisor), Said Beidas (photograph courtesy of Ms. Nava Chen) 164

8.1 At a citrus exhibition in London: The Pardess exhibit, 1930s (photograph courtesy of Ms. Nava Chen) 185

9.1 Jaffa Port—whereof most of Palestine citrus fruit was exported (photograph courtesy of the Israeli Government Press Office, photographer: Zoltan Kluger) 200

9.2 Isaac Rokach (photograph courtesy of Ms. Nava Chen) 205

Tables

3.1 Jewish-Owned Citrus Areas, Early Twentieth Century (in Turkish dunams =0.919 metric dunam) 53

3.2 Jewish Owned Citrus Areas, 1927 and 1936 66

3.3 Geographical Centers of Arab Citriculture, 1935 81

4.1 Net Return and Profitability in Orchard Crops, 1911 102

4.2 Investment per One Dunam of Jewish-Owned Orange Grove, 1913 (five years) 107

4.3 Profit and Loss per Dunam of Productive Jewish Orange Grove, 1913 109

5.1 Profit and Loss per Dunam of Productive Orchard, with Jewish Labor, Arab Labor, and Mixed Labor, 1932 115

5.2 Five-Year Investment per Jewish-Owned Orange Orchard Dunam and Production in Fifth Year, 1913, 1929, 1937 (P£, 1936 prices) 150

5.3 Change in Cost of Cultivating and Preparing One Dunam of Jewish-Owned Orange Orchard in First Five Years—1913 as Base Year, in Comparison with 1929 and 1937 151

5.4 Age/Output per Dunam of Orange Orchard 152

5.5 Current Costs per Case of Oranges, Jewish Grove, 1926, 1938 153

7.1 Current Costs per Crate of Oranges in Jewish Grove—from the Grove to the Foreign Market (in Palestine Mills; 1926, 1938) 171

7.2 Profit and Loss per Dunam of a Jewish-Owned Orange Grove, 1926, 1938 172

7.3 The Main Markets and Their Share in Citrus Exports from Palestine (in percentages) (selected seasons) 176

Preface

Many institutions and individuals at Ben Gurion University of the Negev, my academic home, have facilitated the publication of this book. I would like to thank the Ben-Gurion Research Institute in Sede Boker and its former director, Dr. Tuvia Friling; the Department of Jewish History; the Faculty of Humanities and Social Sciences; the rector's office and rectors Professor Nachum Finger and Professor Jimmy Weinblatt; and the president of Ben Gurion University, Professor Avishay Braverman, for their support.

As I was preparing this book for publication, I was fortunate to have the opportunity to discuss my ideas with colleagues and friends. I am particularly grateful to Nachum T. Gross, Jacob Metzer, Ilan S. Troen, and Derek J. Penslar.

The Ben-Gurion University–University of California, Los Angeles Academic Exchange Program, headed by Prof. Fred Lazin at BGU and Prof. Samuel Aroni at UCLA, enabled me to travel to California, explore its citrus industry, and gain access to important archival material. Special thanks go to Prof. Aroni for his warm hospitality in Los Angeles. I would also like to thank Prof. Stephen Aron from the history department at UCLA and Prof. Teofilo Ruiz, the department chair, for hosting me at UCLA. In his capacity as director of the Institute for the Study of the American West, *Autry National Center*, Prof. Aron also invited me to lecture at the Autry Institute Workshop. I benefited greatly from the stimulating and lively discussion there.

It is a pleasure to thank Dr. Vincent H. Moses, director of the Riverside Municipal Museum, for hosting me at the museum and at the California Citrus State Historic Park. I learned a great deal from our memorable and enjoyable encounter.

While in the States I had the opportunity to study new archival material pertaining to this book. I am grateful to the staff of the Department of Special Collections, Young Research Library, University of California, Los Angeles; the Bancroft Library, University of California, Berkeley; the Department of

Special Collections, Shields Library, University of California at Davis; and the Manuscripts and Archives Division of the New York Public Library.

This volume is an English-language version of my book *Citrus Blossoms: Jewish Entrepreneurship in Palestine, 1890–1939*, published in Hebrew by Magnes Press in 2000. I would like to thank Magnes Press and its former director, Dan Benovici, for their kind cooperation. Many thanks to Naftali Greenwood for his professional and accurate translation. I thank Tamar Soffer from the department of geography at the Hebrew University of Jerusalem for preparing the maps included in this volume. I am grateful to the following individuals and institutions for allowing me to reproduce photographs and visual images from their collections: The Nahum Gutman Museum, Neve Zedek Tel Aviv; Ms. Cila Agam; Ms. Yael Miller; Ms. Nava Chen; Mr. Kevork Kahvedjian of Elia Photo Service, Jerusalem; the Israeli Government Press Office; the Central Zionist Archives; the Lavon Institute for Labor Research; and Ms. Hana Pinshow, head of the Ben Gurion Archives, the Ben Gurion Research Institute.

As always, I dedicate this book to my beloved wife, Gannit, and to our three wonderful children, Lee-Or, Roi, and Amir, with all my love.

ONE

The Tantalizing Aroma of Citrus Blossoms

FOR MOST PEOPLE in the Yishuv—the pre-state Jewish community in Eretz Israel (Palestine)—in the late Ottoman and the British Mandate periods, citriculture evoked a tapestry of images that would later evolve into sacred symbols of the Israeli experience: the tantalizing aroma of citrus blossoms blanketing the whole country, heralding the end of the cold winter and the advent of spring; the picker's pleasant fatigue at sunset after a grueling day in the orchards in agonizing struggle for the "conquest of labor" (the Zionist term for establishing the primacy of Jewish, as opposed to Arab, labor); and the ardent faith, the parched throat, and the clenched fists in vehement protest against Jewish citrus growers in the *moshavot* (privately owned agricultural colonies, sing. *moshava*) who used Arab labor. The imagery included fresh orange juice at streetcorner kiosks and in household kitchens; the reliable low-tech metal orange squeezer with the long handle; the bitter taste of green, unripe oranges, which twists faces into that well-known expression of abhorrence; homemade orange candies and the assorted byproducts of Assis, Ltd. A whole succession of visual symbols identified the citrus enterprise with the young, muscular *haluts* (Zionist pioneer): the sturdy hoe, clutched firmly, the first furrows in the orchard, and the outcome—citrus trees sagging with their bounty of gleaming fruit. All of these were ineradicable images of the burgeoning Jewish settlement project in Palestine.

Underlying this world of symbols was a socioeconomic reality without which it could not have existed. It seems, however, that Israel's historical memory has generally overlooked this reality, just as the inhabitants of the Yishuv banished it from their consciousness. Instead, an alternative value system with its own cohesive set of symbols came into being—a Socialist Zionist worldview that crowded out the capitalist sociocultural reality of Palestine's private Jewish citrus growers and relegated it to the margins of national legitimacy.

From its inception in the second half of the nineteenth century until the outbreak of World War II—which is the chronological framework of this study—Jewish citriculture, based on private capital and hired labor, was a leading industry in Jewish private enterprise in Palestine. However, the vanguard socialists of the Second Aliya (those who reached Palestine in 1904–1914, in the second of five large waves of Jewish immigration), who became the leaders of the Yishuv during the British Mandate era, fashioned a cohesive ideology that considered private capital and private initiative an illegitimate path to the fulfillment of Zionist aspirations. According to the logic of the socialist Zionist leaders, private capital was "anti-national," that is, it aimed not to attain national goals but to amass individual profit. This rationale focused on private citriculture with particular vehemence because this branch of private enterprise offered a viable alternative to the mainstream views of the Zionist Organization (ZO) and, particularly, of the Zionist Labor Movement—both of which had consecrated two basic Zionist values, public ownership of land and Jewish ("Hebrew") labor, as essential conditions for the Jews' national repatriation. We deal with these issues presently, but at this point it suffices to say that private citriculture was unique in the realm of Jewish private enterprise in Palestine, most of which was concentrated in urban settings.

Just the same, the tantalizing aroma of Palestine's orange blossoms was one of the most conspicuous indicators of spring in Palestine for several generations, not only because it was popularized in literature and song but also, and mainly, because it was real. Vast parts of the country were blanketed with citrus orchards, owned by Jews and Arabs alike. Not only did the orchards transform entire regions (such as the Sharon Plain) into settled, cultivated areas, and not only did even cooperative and collective rural settlements affiliated with the official Zionist Movement—moshavim and kibbutzim—join the veteran moshavot in numbering citriculture among their main endeavors in agriculture, but the cities and their peripheries, especially in the coastal lowlands, were soon layered with orchards.

Jaffa, the epicenter of citriculture in Palestine, was a verdant town where visitors since the nineteenth century (and even earlier, in fact) had been presenting detailed and picturesque accounts, including the scents of the orchards and the juicy sweetness of their produce, of the citriculture activity all around. In the early twentieth century and during the British Mandate era (1918–1948), orange groves were still Jaffa's hallmarks. The demographic, economic, and cultural center of the ascendant Yishuv, Tel Aviv—which started out as a neighborhood of Jaffa—was also enveloped in citrus. Even city-bound teenagers who could not see orchards from their windows visited them, if only when they were inducted for harvest work as part of the afore-

FIGURE 1.1. Nahum Gutman, *Orange Groves in Jaffa*, 1928.

mentioned continual struggle for the "conquest of labor." One of the first experiences that Jewish immigrants encountered in Palestine was working with a *turiyya* (hoe) in a "Hebrew" orange grove. Desperate to make a living, the new landed often headed straight from the boat to one of these orchards.

Another recurrent image was the Hebrew worker as the bearer of an ideology. Such people found in their bonebreaking labor in the groves a twofold realization of their Zionist and socialist goals. First, it gave them a supreme opportunity to turn the upside-down "Jewish occupational pyramid" right side up again by "returning to the soil" and acquiring title to that soil by means of grueling physical labor. Second, it allowed them to enlist in the all-out socialist war against the employer, the Jewish farmer, whom they perceived as a ruthless capitalist who lacked national values and concerned himself with the bottom line only.

A typical example of this double-edged attitude toward private citrus growers may be adduced from the Second Aliya pioneer and future leader of the Zionist Labor Movement, Berl Katznelson. A short time after his arrival in Palestine, Katznelson visited Petah Tikva (about ten kilometers northeast of Jaffa) and experienced an epiphany by exhausting himself in manual labor. "It's good for you! Physically good. So good that it can't be expressed in words!" Ironically, twenty-five years later, as a prominent socialist national leader, Katznelson took part in a vigil in the citrus groves of Kfar Saba (twenty kilometers northeast of Tel Aviv) against the erstwhile employers.[1]

Katznelson and other Zionist Labor Movement leaders knew full well that their struggle might be harmful both to the cause of Hebrew labor and to the colonies' socioeconomic vitality. Nevertheless, they preferred to preserve their dichotomous set of values and symbols rather than to endorse the concept of private enterprise in any way. Even a de facto rapprochement between private citrus growers and the Yishuv's Labor leadership during the turbulent years of the Arab uprising (1936–1939) failed to bring about any change in the Labor ideology. Labor's principled attitude toward private enterprise and the citrus industry remained the same—nonrecognition of their contribution to the fostering and solidification of the Zionist enterprise.[2] In view of Labor's ideological hegemony and political dominance of the Yishuv and the State of Israel until 1977, it should come as no surprise that Jewish citriculture was marginalized in public consciousness and in scholarship as an economic and social endeavor that hardly deserved serious attention.

JEWISH CITRICULTURE AS A PRIVATE ENTREPRENEURIAL INDUSTRY

In this respect, the fate of the citrus industry was hardly different from that of other private economic endeavors in the Yishuv. Those endeavors were also crowded out for years—especially in "establishment" historiography—in respect to their contribution to the Jewish settlement venture in Palestine. This outlook, however, has changed recently. Today, Jewish private enterprise in Palestine at the end of the Ottoman period and throughout the British Mandate has gained increased attention.[3] Economic and demographic data show that it was the private sector, not the public sector of the organized Zionist movement, that was dominant in the economic, demographic, and even social development of the Yishuv.[4] Ideology and its attendant rhetoric notwithstanding, the occupational pyramid of Jewish society in the Diaspora was not overturned in the Palestine context; there was no mass outflux of Jewish laborers from cities to villages. The overwhelming majority of Jews in pre–World War II Palestine remained city dwellers, as they had been in their countries of origin.

The composition of the Yishuv's employment structure, however, was markedly different from that in the Diaspora. Diaspora Jewry and its choice of occupations were pronouncedly urban. Among those who moved to Palestine, however, 21 percent of persons employed in 1922–1939 took up agricultural labor, a vocation hardly encountered in the Diaspora. This, however, did not amount to a revolution. Most Jewish labor gravitated to service industries (which employed half of the Jewish labor force in this period) and manufacturing (18 percent in 1922, 23 percent in 1939), two fields of economic activities that were dominated by the private sector. What is more, most capital (80 percent) that flowed into the country between the two world wars was private.

An itemization of the Jewish domestic product in Palestine in 1922–1935 shows that service industries accounted for 65 percent; manufacturing 22 percent, and agriculture only 13 percent (all in average terms for the period). Agricultural settlement, of course, was the ideological and practical jewel in the Labor Movement's crown and the recipient of most of the Zionist Movement's budgetary exertions between the world wars. However, the data show that most Jewish agricultural output derived from citrus (38 percent in 1922 and a whopping 75 percent in 1935), an industry almost exclusively in private hands, and not from the dry farming activities that were typical of the Labor Movement's settlements.[5]

Moreover, from the end of the Ottoman era and especially under the British Mandate, citriculture was the country's leading export industry in both the Arab and Jewish sectors, which divided up harvests and orchard area almost equally. Citrus exports accounted for 43 percent of export's value in 1927 and nearly 84 percent by 1935.[6]

It is a less-known fact that during this period, the Jewish National Fund (the JNF—the land purchase and development agency of the Zionist Organization) held title to only a small fraction of Jewish-owned land. Most Jewish land was owned by private entrepreneurs, many of them citrus growers. This stands in stark contrast to the conventional wisdom about Jewish settlement as a project conducted almost exclusively on "nationally," that is, Zionist Organization, owned land. By 1914, the JNF had managed to acquire a mere 4 percent of Jewish-owned land. In 1929, it held 23 percent of all areas in Jewish hands. About 31 percent was owned by the Palestine Jewish Colonization Association (PICA, a society that promoted Jewish settlement in Palestine, established by Baron Edmond de Rothschild), and the remainder, approximately 46 percent, was held by private owners. Even at the end of 1936, only 26 percent of Jewish land was under "national" ownership.

Most JNF-owned land in this period, roughly 74 percent, was situated in three valleys in the northern part of the country: Zevulun and Jezreel (48 percent) and Jordan (26 percent). The remainder, owned by private individuals and the PICA, was on the coastal plain—from Zikhron Yaakov and Hadera in the north to Nes Tsiyyona and Rehovot in the south. In the interwar period, this was the country's main citriculture region.

Agricultural settlement financed by the ZO institutions lacked a majority in additional respects. As of 1936, only eighty-five of 203 Jewish settlements (42 percent) had been established by the "national" funds (Keren

Hayesod and JNF). Of the 118 other localities, seventy-seven had been settled by private companies and individuals; twenty-two by PICA; and seventeen by various public (but not "national") companies.

The demographic picture is even clearer. Only 23 percent of the Jewish rural population dwelled in "national" (Keren Hayesod) settlements as of the end of 1936. The vast majority of Jewish settlers, 77 percent (76,000 of 98,000), lived in privately owned agricultural villages—moshavot—where the main type of farming was most often citriculture.[7]

Due to the size of the Jewish citrus industry in "dunamage" (the land unit during this time, and to this day in Israel, was the *dunam*—one-tenth of a hectare and approx. one-fourth of an acre) and number of orchard owners, and due to its importance as the Yishuv's main export industry, citriculture provided a livelihood for tens of thousands in occupations as diverse as unskilled picking, carpentry and haulage, and expert irrigation and entomology. In times of economic crisis, for instance, the one that erupted during the Fourth Aliya (1924–1928), citriculture provided employment for thousands of idled urban workers. Its growth led to economic, social, and demographic expansion and the consolidation of the veteran moshavot, now termed "orchard moshavot," and in new localities that were based on citrus *ab initio*. This consolidation facilitated the absorption of thousands of immigrants and gave many of them an opportunity to experience the Zionist "rite of passage" of physical labor and return to the soil.

Nevertheless, in the contest for the Israeli collective memory the "socialist valley" defeated the "capitalist coastal plain." Wheat fields took precedence over citrus groves, pioneers vanquished orange growers, and the ideological struggle for Hebrew labor defeated the burgeoning cooperation between Palestinian-Arab citriculture (and citriculturists) and Jewish citriculture (and citriculturists).

It is not the intention of this book to deal with the enshrinement of these images in the Israeli collective memory; this is a subject that requires separate research. Instead, the following chapters will sketch a realistic approach for the examination of their empirical roots that, like all images, were anchored in sociocultural contexts. Additionally, this study does not intend to disparage the important role played by the Zionist Labor Movement and, *a fortiori*, the Zionist Movement, in realization of the Zionist enterprise. The history of the Yishuv and the Zionist settlement venture simply cannot be described if the Labor Movement and its crucial role are overlooked. Instead, my purpose here is to probe and evaluate the nature and activities of private enterprise, an important sector of the Yishuv that has been given short shrift in research and public consciousness.

Thus, the main goal of this book is to examine the ideology and the socioeconomic activity of private enterprise in the Yishuv through the prism of Jewish citriculture. In view of the dominant étatist and socialist ideological views of the Zionist leaders, who wished to subordinate Jewish private initiative to the agenda of their movement and its institutions, I attempt to examine how the private citrus growers explained their worldview and their economic activities. In many respects, Jewish citriculture may serve as a model for the status that private enterprise attained in the Yishuv during its lengthy formative period between the First Aliya (1882–1904) and the beginning of World War II, in its ideological rationales and its economic modus operandi. Just the same, the citrus industry was also unique among areas of private enterprise such as services, manufacturing, and construction, due to its activity in the major ideological and symbolic arena that the Zionist Movement claimed as its own: agricultural settlement, return to the soil, acquisition of land, "conquest of labor," and relations with the local Arab population. This seems to lend the study of Jewish citriculture as a settlement venture and as a paragon of private enterprise a much broader meaning than that of other private enterprises.

In addition to the main goal noted above, our treatment of citriculture will diverge from the accepted macroeconomic point of view in the economic discussion of Jewish Palestine. Thus, the economic analysis in this study relegates aggregate indicators (such as product or output at the national level) to the background and strives to show how basic economic considerations at the individual or firm level shaped activity in the industry at large.[8]

FOUR SPHERES OF INFLUENCE

Four main spheres of influence created a framework within which Jewish citriculture functioned. The first encompasses the nature and modus operandi of *Zionist settlement as a form of European colonial settlement.* It was a settler society but a unique one, driven mainly by modern national motives, with universal features rooted in the Judeo-Christian ethos and a special emphasis on the repatriation of the Jewish people. The overall nature of Jewish national settlement in Palestine as a colonial endeavor is not the concern of this study; research wages a lively debate about the matter.[9] Nevertheless, there can be no doubt that in the first sixty years of Zionist colonization in Palestine beginning in the early 1880s, an economically and socially developed European immigrant population settled an underdeveloped non-European region inhabited by an indigenous population. During this settlement process, land was acquired from local inhabitants to create a territorial basis for the establishment of a European society alongside the indigenous one. Furthermore, the institutional and conceptual framework of European colonialism, especially before World War I—one that included the formation of settlement companies, financial services, and metropolis-colony relations—was the one in which most Jewish immigrants, Labor Movement pioneers and private settlers alike, actualized the process of their return to the Land of the Patriarchs. After the British conquest, the country came under the dominion of the world's foremost colonial power, and the regime set up in Palestine was crafted and administered under the guidelines and outlooks of a British crown colony. What is more, Britain was obligated under the terms of its Mandate from the League of Nations to implement a pro-Zionist policy that would facilitate the realization of the Zionists' goal.[10]

From the inception of the British Mandate to the beginning of World War II, a fully integrated, modern Jewish national society was established in the country. The local Arab population, however, did not simply vanish, and two interacting national societies developed side by side.

This process—colonial settlement at the outset and the formation of a dual society later on—also affected Jewish citriculture. The availability of capital, land, and labor were influenced by the nature of Jewish settlement as a colonial endeavor and this, in turn, greatly influenced the citrus growers' social and economic decisions. Moreover, the new Jewish settlement project was a distinctly modern element in the premodern social and economic context of late-nineteenth-century Palestine. Even in the first half of the twentieth century, during the Mandate era, the Yishuv remained the leading element of modernity despite signs of growth and modernization in the Arab economy.[11] Thus, the modern dimension of the Yishuv economy had a considerable effect on the nature of Jewish citriculture. Finally, political relations with the Arab population made an economic impact on Jewish citrus growing, as discussed in the following chapters where necessary.

The second sphere of influence was that of *interrelations with Arab citriculture in technology, organization, financing, and marketing*. These matters are discussed below, but here it should be emphasized that citriculture in Palestine was not a Jewish import; it was already considerably developed at the time of the Jews' arrival in the late nineteenth century and had been exporting produce to Britain, Turkey, Egypt, and other countries. Arab citriculture was concentrated in the vicinity of Jaffa, close to the country's main port, where climatic, agronomic (especially in the availability of water), and marketing conditions were well suited for its development.

When the Zionist settlers appeared, only Arab-owned plantations produced citrus for export. In 1914, Arab citriculture still accounted for more than 72 percent of the country's citrus exports. Even during most of the interwar period, Arab citrus exports surpassed that of the Jewish industry. Not until the mid-1930s did the Jews catch up. Arab growers did as much new planting as Jewish growers. By the end of the period at issue, Jewish and Arab citrus holdings were equally divided and Jewish exports did not account for more than 60 percent of total citrus exports. Thus, Arab citriculture also expanded significantly and, like Jewish citriculture, gave evidence of vigorous capitalistic activity.

The history of Arab citriculture is not the main subject of this book and data on the Arab sector is still fragmentary, with much room left for specialized research that entails access to Arab sources from the era in question. Such research would explain in greater detail the role of Palestinian-Arab citriculture in the industry at large. Be this as it may, the Arab citrus industry "accompanied" the Jewish industry during the late Ottoman and interwar periods, when Arab growers worked in rather close contact with their Jewish counterparts. Therefore, the Arab sector will be a permanent object of reference in this study and, where possible, the discussion will deal specifically with various aspects of the Arab growers' activities.

The third sphere of Hebrew citrus growing is connected to *the negative attitude of the Labor Movement* toward private agriculture in particular and to private enterprise in large in the context of the national economy. This challenge appeared as a permanent cultural feature throughout the emerging Zionist society, and its influence on private citriculture will be discussed in chapter 2.

The fourth sphere includes the policies and attitudes of the *Mandatory Government* in Palestine and H.M.G. in Britain. In stark contrast to the mainstream étatist and socialist ideological outlook of the Yishuv, which disparaged and criticized private economic and social activity, the Mandatory Government lent it full support. The government's stance, flowing from the worldview of economic liberalism, severely limited its own involvement in the economic life of the country generally and in citriculture, a bastion of private enterprise, in particular. Concurrently, the government assumed the "classic" economic functions of the liberal school by facilitating the creation of conditions for private activity, for example, establishing a monetary system, providing a transport and communications infrastructure, and so on.[12]

These four broad spheres of influence, within which Jewish citriculture operated, will serve us as a comprehensive frame of reference that will reside in the background of our discussion and, where necessary, will receive our attention.

THE ENTREPRENEUR AND ENTREPRENEURSHIP: THEORETICAL CONTOURS

Jewish citriculture during the Mandate era was a private enterprise or "entrepreneurship," and the person engaged in this activity was an "entrepreneur." Notably, scholarly opinion is divided about the precise definitions of those terms. One of the most widely accepted definitions of an entrepreneur—a word coined in 1730 by Richard Cantillon, an Irish banker residing in France—is a businessperson who operates and is willing to take risks under conditions of uncertainty. According to classical theoreticians of economics, however, as Yair Aharoni notes, the economic system contains a known number of variables about which there is complete information and certainty; thus, in their view, the entrepreneur actually plays a rather limited role. In the economic systems of the classical theoreticians, the individual who acts within an economy is a "capitalist." However, these thinkers fail to determine whether the capitalist's "profit" is the price of the use of capital or whether the burden of uncertainty is part of it as well.

Today, the accepted view among economists is that the entrepreneur fulfils a vital role in the economic system, especially in its development. In 1921, Frank Knight, like Cantillon, defined the entrepreneur as a person who operates under conditions of "uncertainty" and, therefore, faces the hazard of immeasurable risk. Such activity stands in sharp contrast to the task of managers, who function under conditions of certainty and measurable risk.[13]

The most important theoretician in the matter of defining entrepreneurship, Joseph Schumpeter, also notes the "entrepreneur"–"manager" dichotomy. Schumpeter argues that the entrepreneur not only bears the risk and operates under conditions of uncertainty but is also primarily an innovator and trailblazer in technologies, discovery of undeveloped markets, and more efficient organization of existing resources. The entrepreneur is characterized by profoundly operating outside the area of the routine. Schumpeter stresses the importance of the entrepreneur as the bearer of technological and economic progress, an innovator who disrupts the existing equilibrium in the capitalist system and, by so doing, generates short-term cycles of instability and a long-term process of growth. The manager, in contrast, remains part of the static system.[14]

Schumpeter's critics accuse him of investing the "heroic" nature of the entrepreneur with excessive importance and belittling processes of adjustment and change amidst routine activity. Thus, another theoretician, Fritz Redlich, states that when economic reality is observed not from the "heroic" point of view but from a long-term perspective, it turns out that "subjective innovation is the routine" in all successful business activity. Redlich stresses processes of adoption and adaptation, rather than invention, in the transfer of current technologies; the innovation that occurred in certain family firms for generations, and the range of activities that a businessperson carries out as manager, entrepreneur, and innovator. However, Redlich attempts to distinguish between manager and entrepreneur not only in firms where the manager is also the owner but also in large corporations that are administered by a complex hierarchical system. He sums up his argument by defining an entrepreneur as the player who makes strategic economic decisions, be it the owner of a private company or the senior management of a corporation.[15]

David Landes defines an entrepreneur as "a decision-maker of the economy." Landes's definition includes small business owners, "the newer class of pure managers" (a distinction in which he was preceded by Redlich), and decision makers in government bureaucracies.[16]

Another characteristic of the entrepreneur was observed by Arcadius Kahan, who also recognized the elements of risk and innovation in entrepreneurial ventures. Kahan's unique contribution to the theoretical definition of the entrepreneur was his insight that the entrepreneur recognizes genuine opportunities that others fail to see.[17] Several years ago, Peter Temin, returning to the classic definitions of Knight and Schumpeter, made an effort to clarify the definitions of these terms. He claimed that entrepreneurs are "agents of change" who tackle the challenge of the unknown and "unmeasurable risk." Managers, in contrast, are "agents of stability" who deal with "measurable risk." An important observation by Temin, which takes Redlich's view one step forward, is that in a modern economy, in which large corporations are important players, the manager—who is generally not the owner—becomes the entrepreneur. Although Temin's reasoning does not challenge Redlich's frontally, his observation implies that it does not suffice to make strategic decisions in economic matters, as Redlich postulates, since such decisions may also be made by the manager, whose level of activity is stable. For business decisions to be considered entrepreneurial, they should lead to changes and meaningful breakthroughs.[18]

In addition to predominantly economic definitions of entrepreneurial initiative, there are observations that pertain to the degree of legitimacy that society grants to such activity. Legitimacy (or lack of it) may be spelled out within a defined ideological system or may exist within a value system that is not always defined in explicit ideological terms. Yair Aharoni, for example, who defines an entrepreneur as one who is willing to accept the consequences of economic activity under conditions of uncertainty, states that the degree of risk the entrepreneur assumes is subjective and depends largely on society's attitude toward entrepreneurial activity. Alexander Gerschenkron, in contrast, finds that negative social attitudes toward private enterprise have no significant effect on economic growth and initiative unless they are accompanied by government sponsored anti-entrepreneurial actions.[19]

Additional noneconomic factors that have been found to influence private entrepreneurial activity are characterized by conflicting tendencies. For example, social mobility spurs private enterprise but its absence may also trigger private initiative. The entrepreneur's social status has the same clashing effects. On the one hand, marginal status or location often triggers private enterprise, as the entrepreneur is eager to integrate into the mainstream and to exploit economic opportunities that the mainstream ignores. This posture is often stressed in explaining the entrepreneurial activities of immigrant and ethnic groups.[20] On the other hand, extreme social marginality is unwanted because it distances those on the fringes from the institutions of society at large. Thus, research on the noneconomic factors in entrepreneurship emphasizes social integration, rather than extreme marginality, as a prerequisite for entrepreneurial activity.

David McClelland and Everett Hagen focus on the psychological element in entrepreneurship. McClelland notes the private entrepreneur's need for achievement and Hagen regards the individual's or group's loss of social status as a motivating factor. Paul Wilken, in contrast, analyzes six leading industrialized countries (Britain, France, Germany, Japan, Russia, and the United States) and finds that, among noneconomic factors, psychological factors and social mobility have no perceptible effect on entrepreneurial activity, whereas ideology and the degree of legitimacy accorded by society to private enterprise have a positive influence. Notwithstanding this, according to Wilken, the noneconomic factors were not the ones of consequence in determining the extent of entrepreneurial activity.[21]

Thus, research seems to offer too many observations about entrepreneurs and entrepreneurship, and the factors that influence them, to determine which of them is "correct." Nevertheless, we will use these definitions in our study on citriculture in Palestine. In our summary, we will ask how well these models fit the entrepreneurial endeavors of the private Jewish citrus growers.

One more observation, related to this study, should be discussed: the distinction between private and public enterprise. The public sector is usually identified with governmental or quasi-governmental organizations, that is, it is defined not as an aggregate of individuals who maintain absolute autonomy within this body but as a separate legal and social entity that works actively on behalf of all individuals. That is, it acts for the common amorphous entity known as "the public at large." Thus, the express purpose of public (or quasi-public) entrepreneurial activity is the public's welfare, whereas private enterprise aims to benefit the individual investor. There is considerable truth to the claim that in the modern economy, in which company or corporation ownership is shared by many individuals, the dividing line between public and private ownership of capital has become blurred. Even business activity now operates in a "mixed" condition and not according to traditional dichotomous distinctions between the private and the public spheres. However, even those who claim that the "mixed" condition in the modern economy is the most pervasive one acknowledge a quantitative and qualitative difference between public and private enterprise.[22]

The following discussion will draw on this distinction in differentiating between the entrepreneurial activities of private Jewish growers in Palestine and those of the Zionist public (or quasi-public) sector and, especially, the Labor Movement.

JEWISH CITRICULTURE: GENERAL PROFILE

Jewish citriculture, as stated, was largely a capitalist activity in which private capital and hired labor combined to attain maximum profit in a market economy. Profit was the overarching interest of Jewish plantation owners, both individuals and companies; it was also the major influence on the rapid growth of the industry.

Nearly all citrus fruit grown in Palestine was shipped to foreign markets, foremost England and elsewhere in Europe. After World War I, citrus exports to Europe dramatically increased, mostly due to the healthful qualities related to the fruit and a change in consumers' preferences in favor of fruit. The growing demand generated a price spiral that caused profits to skyrocket.

The high profitability of citriculture prompted many private entrepreneurs—in Palestine and abroad—to invest in new plantations. Despite the long maturation period (five to six years until first harvest and nearly ten years until full harvest), private entrepreneurs—Jews and Arabs alike—continued pouring money into the industry throughout most of the interwar period.

According to accepted estimates, there were approximately 29,000 dunams of citrus orchards in Palestine at the onset of Mandatory rule (1920), of which 10,000 dunams (35 percent) were Jewish owned. Twenty years later, in 1940, the planted area had expanded to 300,000 dunams, of which 155,000 dunams (52 percent) were in Jewish hands. Thus, during that time, Jewish-owned plantations grew by a factor of 10.3. Citrus exports also surged, from 830,000 cases in 1920/21 to 11.5 million cases in 1937/38—a fourteenfold increase. Between 1927 and 1931, Palestine was the world's fourth-largest citrus exporter (after Spain, the United States, and Italy); ten years later it reached second place (after Spain).

The dramatic expansion of Jewish citriculture was due chiefly to the profitability (actual and expected) of this industry. From this perspective, the period at issue may be divided into four main subperiods:

1. 1890–1919—inception and expansion of the Jewish industry. The latter part of this period, 1914–1918, coincides with the debilitating crisis of World War I.
2. 1920–1929—the postwar period of rehabilitation and high profitability.
3. 1930–1934—years of reasonable profitability.
4. 1935–1939—severe crisis and steep decline in profitability.

There was a clear correlation between the profitability of the industry and its expansion, as reflected in the extent of new plantings. In this respect, the period may divided into five subperiods:

1. 1890–1914—the birth of the Jewish citrus industry in Petah Tikva (ten kilometers northeast of Tel Aviv) and its introduction in the Jewish colonies.
2. 1921–1925—the postwar rehabilitation, during which much less new planting was done than in subsequent years. Only 1,500 dunams were planted (evidently by Jews in greater part).

IV. Census of Jewish Citriculture in Six Plantation Blocks, at the End of 1936

IV. מפקד הפרדסנות העברית בששה גושי מטעים, בסוף שנת 1936

Table 68. Summary Figures of Area of Citrus Plantations in Palestine and Export of Citrus Fruit during 1922 — 1938
(according to the records of the Govt., Jewish Agency and various sources)

טבלא 68. סיכומים כלליים של שטח מטעי ההדר בארץ ואכספורט פירותיו בשנים 1922 — 1938 (לפי מספרי הממשלה, הסוכנות היהודית ומקורות אחרים)

בזה מפרדסים של יהודים Thereof, from Jewish groves	Export of citrus fruit from Jewish and Non-Jewish groves (according to seasons) אכספורט פרי הדר מפרדסים של יהודים ולא יהודים (לפי עונות): Total בסך־הכל — Value in LP. ערך בלא״י	Total בסך־הכל — Cases מספר התיבות	Lemons, etc. לימונים ושאר מטעי הדר — Value in LP. ערך בלא״י	Lemons, etc. — Cases מספר התיבות	Grapefruit אשכוליות — Value in LP. ערך בלא״י	Grapefruit — Cases מספר התיבות	Oranges תפוחי זהב — Value in LP. ערך בלא״י	Oranges — Cases מספר התיבות	Season עונה 1.IV—31.V	Export, in cases, to principal countries of destination (according to calendar years) 1.I — 31.XII כמות האכספורט לארצות הקליטה העיקריות, בתיבות (לפי השנה האזרחית): Rumania לרומניה	Poland לפולין	Holland להולנד	Germany לגרמניה	United Kingdom לאנגליה	Area of citrus plantations in metric dunams שטח כמעי הדר בדונמים מטריים: Total בסך־הכל — Cumulative to end of year סך־הכל מטעים בסוף השנה	Total — Planted annualy הנטיעה בכל שנה	Non-Jewish owners של לא יהודים — Cumulative to end of year סך־הכל מטעים בסוף השנה	Non-Jewish owners — Planted annually הנטיעה בכל שנה	Jewish owners של יהודים — Cumulative to end of year סך־הכל מטעים בסוף השנה	Jewish owners — Planted annually הנטיעה בכל שנה	Year השנה
22	21	20	19	18	17	16	15	14	13	12	11	10	9	8	7	6	5	4	3	2	1
%							205 615	830 959	1920/21					(330 000)							1921
							333 716	1 234 251	1921/22			19 909	79	743 179	29 000		19 000		10 000		1922
							420 054	1 365 543	1922/23	2 332		364	8	942 463	29 500	500	19 000		10 500	500	1923
							431 582	1 589 331	1923/24	18 821		—	1 305	1 190 473	30 000	500	19 000		11 000	500	1924
							621 276	2 146 457	1924/25	17 794		915	4 169	1 319 667	30 500	500	19 000		11 500	500	1925
	467 632	1 518 731	969	3 615	—	—	466 669	1 515 116	1925/26	16 499		500	22 304	1 567 301	38 500	8 000	21 500	2 500	17 000	5 500	1926
	827 897	2 668 291	2 851	9 575	—	—	825 046	2 658 716	1926/27	36 729		13 569	111 678	1 943 828	56 000	17 500	32 000	10 500	24 000	7 000	1927
	654 820	2 220 443	2 687	10 135	—	—	652 133	2 210 308	1927/28	59 965		9 406	143 978	1 730 520	67 000	11 000	36 000	4 000	31 000	7 000	1928
	539 512	1 802 547	3 695	12 789	930	2 265	534 887	1 787 493	1928/29	40 579	350	39 582	245 681	1 333 674	87 000	20 000	41 000	5 000	46 000	15 000	1929
	784 777	2 610 205	1 809	6 333	5 622	13 011	777 256	2 590 861	1929/30	41 278	2 197	27 162	502 009	2 167 897	106 500	19 500	46 500	5 500	60 000	14 000	1930
37%	744 513	2 469 856	1 264	4 803	15 602	39 938	727 647	2 425 115	1930/31	49 677	11 207	16 778	551 862	1 901 510	121 500	15 000	51 500	5 000	70 000	10 000	1931
37 "	1 782 261	3 698 489	2 540	7 729	57 569	105 811	1 725 152	3 584 949	1931/32	101 957	14 053	51 166	678 463	2 620 966	156 500	35 000	66 500	15 000	90 000	20 000	1932
48 "	2 097 393	4 490 409	6 949	16 261	129 444	244 603	1 961 000	4 229 545	1932/33	79 529	22 555	87 147	761 427	3 230 290	200 000	43 500	80 000	13 500	120 000	30 000	1933
51 "	2 633 380	5 533 350	8 318	22 323	183 584	353 250	2 441 478	5 157 777	1933/34	122 922	45 070	150 155	1 206 772	3 621 538	250 000	50 000	105 000	25 000	145 000	25 000	1934
58 "	3 382 964	7 334 343	12 062	26 435	341 585	682 857	3 029 317	6 625 051	1934/35	176 812	233 704	387 607	518 391	5 550 125	278 000	28 000	125 000	20 000	153 000	8 000	1935
60 "	2 535 870	5 886 401	31 901	50 336	304 987	843 811	2 198 982	4 992 254	1935/36	146 668	248 226	392 636	286 209	4 505 914	298 000	20 000	143 000	18 000	155 000	2 000	1936
60 "	3 873 429	10 795 894	29 760	71 501	465 824	1 533 710	3 377 845	9 190 683	1936/37	241 871	366 405	800 605	261 894	8 390 097	299 500						1937
60 "	3 908 262	11 494 079	31 200	80 281	502 330	1 800 807	3 374 672	9 612 901	1937/38						300 000						1938

FIGURE 1.2. Gurevich and Gerz, "Area of Citrus Plantations and Export Figures, 1922–1938."

3. 1926–1931—the "planting frenzy" period, at nearly 91,000 dunams (of which 58,500 dunams, 64 percent, were owned by Jews). By the end of the subperiod, the new plantations accounted for 30 percent of all citrus areas in 1939.
4. 1932–1936—the heyday of new planting. Within five years 176,500 dunams were planted (59 percent of all citrus area in 1939), of which 85,000 dunams (48 percent) belonged to Jews.
5. 1937–1939—a drastic reduction in planting, almost to the point of complete cessation.[23]

The growing subperiods do not always coincide perfectly with the profitability subperiods because the response to changes in the level of profitability, as reflected in the extent of new plantings, came at a lag and because variables not directly linked to profitability were also at work. These variables included the global economic depression, the growth of the Yishuv, and the amount of capital seeking investment opportunities. There seems, however, to be a direct and clear correlation between profitability (and expectations of profitability) and the extent of planting, which exceeded all forecasts. The third subperiod (the "planting frenzy") began at the height of the severe economic crisis during the Fourth Aliya (1924–1928); at this time, citrus planting succeeded in mitigating the harshness of the crisis and gave private entrepreneurs access to cheap labor. The fourth subperiod basically overlapped the boom of the Fifth Aliya (1930–1939), when the country enjoyed a healthy inflow of private capital but wages and land costs rose. These background factors definitely had some effect on the correspondence between new planting and profitability, as we show below. These scanty indicators confirm the importance and centrality of profit (and expectations of profit) in the expansion of Palestine citriculture. Additional factors, such as the Zionist worldview and the aspiration to contribute to the country's development, also played a considerable role in investment in Jewish citriculture. The critical factor in determining the economic behavior of Jewish citrus growers in the interwar period, however, remained the profit motive.

World War II ended the period when the Palestine citrus industry could operate with almost unrestrained license. When the war broke out, the country's economy shifted to a closely regulated war footing. In late 1940, the Mandatory Government established the Citrus Control Board and the Citrus Marketing Board, which brought the Arab and the Jewish sectors under one umbrella and within one regulatory framework. Thus, 1940—and in many respects September 1939—serves as a convenient *terminus ad quem* for our discussion of Jewish citriculture as an arena of unrestrained private entrepreneurship in the pre-Israel era.

Many sources were used in this study. They include archival material—chiefly from the Central Zionist Archives, the Israel State Archives, the

Farmers' Federation Archives, the archives of early Jewish settlements, and private archives—and many contemporary publications. Numerous detailed articles on the industry appeared in the Hebrew press of the time, especially *Hadar* ("Citrus"), a monthly on citrus matters that began to circulate in 1928, but also *Bustenai* ("The Gardener"), the journal of the Farmers' Federation, whose members were mainly citrus growers. Other journals cited include as *Mis'har ve-ta'asiya* ("Trade and Industry"), *Ha-haklai* ("The Agriculturist"), and *Ha-sadeh* ("The Field"). General and economic writings in the Hebrew press and special publications of the government's Agricultural Department were used in the research, as were reports by experts whom the Mandatory Government and the Farmers' Federation invited to the country. The memoirs of, and in several cases interviews with, contemporaries of the time were invaluable in my research.

COMMONLY USED CONCEPTS AND TERMS

The standard botanical classification of citrus fruit places the genus citrus in the *rutaceae* family. The Hebrew name of the citrus is *hadar*. *Hadar* is made up of sixteen species, eight of which are well known and cultivated: citron (etrog), lemon, "sweet lemon," sour orange, mandarin, shaddock, grapefruit, and sweet orange. The last-mentioned includes several well-known varieties; foremost among them in Palestine was (and still is) the Shamouti.

The source of the name Shamouti is not clear. The usual explanation attributes it to a local tradition based on the Arab word for the oval oil lamps that were common in Jaffa in the late nineteenth century, when this variety of orange began to appear on a large scale.[24]

The Shamouti's uniqueness relative to other varieties lies in its handsome shape, delicious flavor, aromatic fragrance, paucity of seeds, and a thick, cushion-like peel that enables it to withstand jolts in transportation. During the time discussed in this study, some countries made strenuous efforts to acclimatize the Shamouti in their soil but did not succeed. This only reinforced the singularity of the Palestine fruit in contrast to other varieties.

The Shamouti, known by its trade name of "Jaffa orange" in the late nineteenth century, blossoms between March and April and begins to ripen in late October or early November. In the period of concern in this book, it was still necessary to market the fruit as it ripened. Marketing lasted until April but most of the harvest was shipped between December and March. The export of other citrus fruits, such as grapefruits and lemons, coincided for the most part with the shipping season of the Shamouti. The Jaffa orange accounted for a vast majority of Palestine's citrus exports; at the beginning of the period, hardly any other species or variety of citrus was known in the country. By the end of the period (at the beginning of World War II), however, citrus exports had become somewhat more diverse; according to some

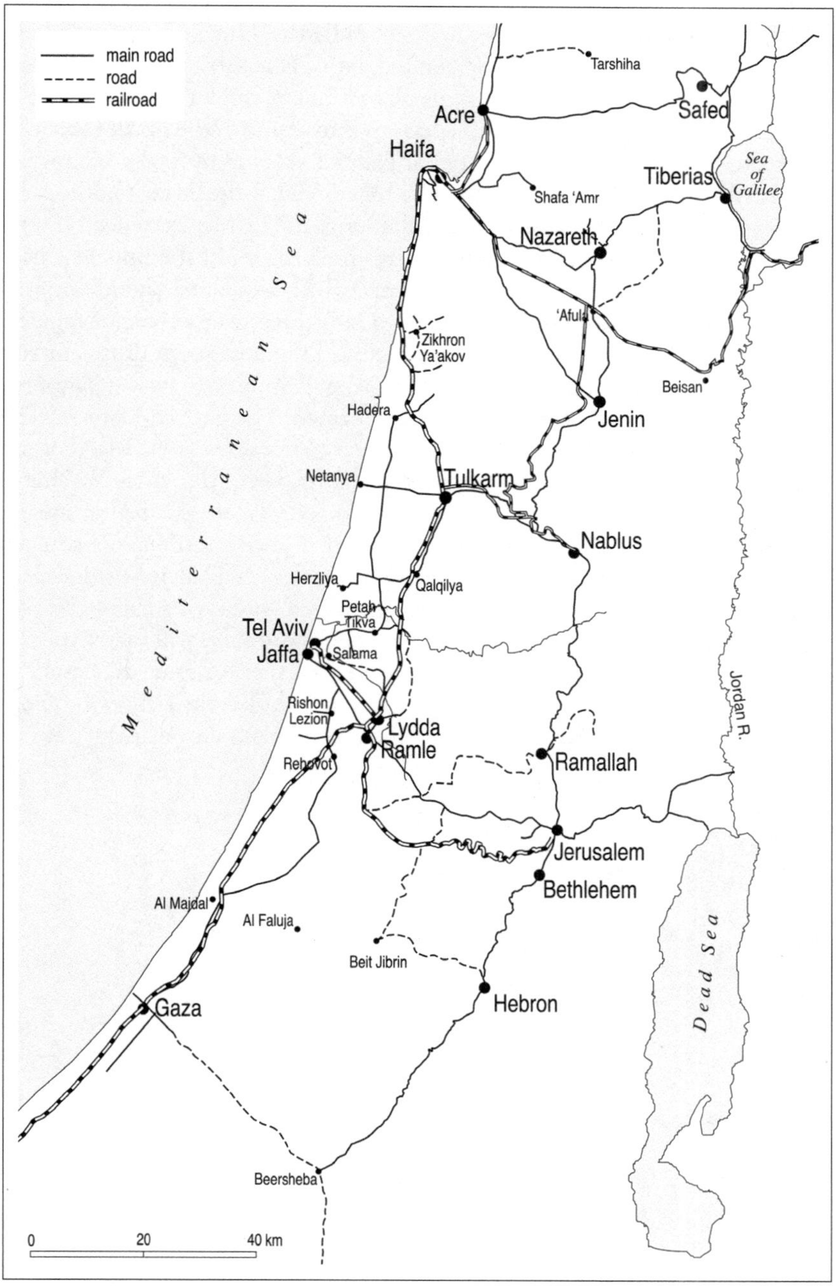

Map 1.1. Palestine in 1936.

estimates, 12 percent of citriculture at that time consisted of grapefruit and 3 percent were lemons, tangerines, and so on. By that time, 85 percent of citrus orchard floor in Palestine was planted with Shamouti.

During the period at issue, citrus plantations covered the greater part of the coastal plain, from Gaza in the south to Binyamina in the north (about 35 kilometers south of Haifa). Additional planted areas were in the vicinity of Acre (10 kilometers north of Haifa), the Jezreel Valley, the Sea of Galilee area, the Jordan valley south of Beit Shean, and Jericho. The Mediterranean coastal plain, however, remained the hub of citrus production. In the linguistic geographical terminology of the Jewish settlers, which was based on and inspired by the Bible, the coastal plain was divided into three main subregions: Judea, the Sharon Plain, and Samaria. Judea included the area south of the Yarkon River where the veteran colonies (Petah Tikva, Rishon Lezion, Nes Tsiyyona, and Rehovot) and other settlements were located. The Sharon Plain was the region extending north from the Yarkon River to the town of Hadera (but not including Hadera). The largest settlements there were Kfar Saba, Ra'anana, Herzliya, and Netanya, which, along with others, were established in greater part during the Fourth Aliya. Samaria included Jewish settlements situated between Hadera and Haifa. The most prominent were Hadera, Pardes Hannah, Zikhron Yaakov, Karkur, and Binyamina. In the following pages, the geographical definitions of citrus areas and of Palestine at large will follow the terminology of the period. Notably, however, the current geographical definitions of Judea and Samaria, especially those favored by right-wing Zionists, refer to the West Bank and not to the coastal plain (see maps).

PART I

Ideological Platform

TWO

Degania or Petah Tikva?

Private Enterprise in the Worldview of Jewish Citrus Growers in Palestine and Their Opponents

A BROAD IDEOLOGICAL CONSENSUS

THE COMMUNITY OF PRIVATE citrus growers in Palestine was quite diverse and its social makeup changed perceptibly during the lengthy period investigated. The number of players in the industry climbed from a handful in Petah Tikva in the late nineteenth century to several thousand by the beginning of World War II. This aspect of the growers' social profile is the focal point of the next chapter, but what we have said thus far suffices to show how difficult it is to sketch any single ideological framework that can contain all views of a collectivity as socially diverse as this. However, by surveying the literature of the time, the minutes of closed meetings, personal journals, memoirs, and press articles, we find—much as Alexis de Tocqueville found in American society in his famous exploratory visit to the United States—that the growers had formed a broad ideological consensus on the role of private capital and private settlement activity in the building of the Yishuv.[1] Few growers articulated it in explicit and definite terms, but in times of controversy and crisis the sources of the time reflect a broad consensus on the topic. Accordingly, it would probably be correct to state that most growers accepted the principled framework of the consensus. Furthermore, much as Tocqueville found in the United States, the ideological consensus among the growers had simple and definite outlines. They rested on two main principles: national consciousness and a basic capitalistic worldview.

One should differentiate between these fundamentals. In regard to the capitalistic worldview, the Jewish growers who asserted the capitalist principle in their operations obviously had a shared ideological stake in the centrality of this outlook. In social terms, too, one may clearly define them—by way of negation—as people who did not belong to the Labor Movement sector, which also participated in citriculture.

When we ask the same question about the national fundamental in the private citrus growers' consciousness, however, we obtain an equivocal answer. Several indicators, however, may shed light on the private growers' national consciousness.

First, it should be emphasized that the citrus industry of Palestine was centered during the period at issue, in social, economic, and symbolic-value terms, in the plantation *moshavot* ("colonies") that had been established in the Ottoman era. Thus, the "center" of the industry had come into being while European Jews who wished to emigrate still had alternatives to Palestine, chiefly the United States. Presumably, then, the national dimension was meaningful for most growers and their offspring who chose to settle in Palestine.

Second, many new growers who reached Palestine during and after the Fourth Aliya period, after the United States stopped allowing unrestricted immigration, had a pronouncedly national outlook. The citriculture that these immigrants established, in the Sharon area, strongly championed the Zionist cause of "Hebrew labor"—a broadly accepted sign of "national" worldview.

Third, although these farmers did not join the mainstream of the "organized Yishuv"—which was led by the Labor Movement—they never truly dissociated themselves from it, as Horowitz and Lissak have shown. Indeed, the reports of the Farmers' Federation and the minutes of its meetings indicate that the farmers' leading stratum was decidedly nationalistic.[2]

Fourth, many of the growers and their leaders—Moshe Smilansky, Shmuel Tolkowsky, Gad Machnes, and Haim Ariav, among many others—played active roles in an emphatically national institution, in which membership clearly denoted an affiliation with the Zionist collective: the Hagana, the Yishuv's mainstream paramilitary organization. Growers also figured very importantly in the merger of "Irgun B" with the Hagana in 1937 and backed David Ben-Gurion's acceptance of the Peel Commission plan for the partitioning of Palestine (1937). It was probably the last-mentioned factor—and possibly the one preceding it—that prompted Ben-Gurion, the chairman of the Jewish Agency, to visit the general convention of the growers' federation that year (an action that puzzled many) and to bestow on the growers the privileged title of *halutsim* (Zionist pioneers).[3]

Thus, it seems fair to say that most growers, or at least their leading echelon, were of the Zionist national persuasion and that they, like many others, became stauncher in this view as the conflict with the Arabs intensified.

Importantly, the growers entertained considerable disagreements on other questions of concern in the Yishuv. For example, many in the Sharon area moshavot and in important veteran citriculture centers such as Petah Tikva, championed and attempted to implement the idea of "Hebrew labor." By inference, there was no identity between the Hebrew labor idea and a noncapitalistic attitude toward the development of the Yishuv. In contrast, even some members of the Zionist Labor Movement occasionally challenged the principle of "100 percent Hebrew labor," as Anita Shapira has shown. The growers were also of many minds about how to resolve the conflict with the country's Arab inhabitants. Smilansky favored a moderate approach and attempted to find a solution that would allow the national groups in Palestine to coexist. Many others, however, advocated separation and disengagement along the lines promoted by the Labor and Revisionist movements. These issues and others, such as political, social, and intellectual life in the moshavot and among the growers, are outside the purview of this book and are not discussed in it. What matters here is the broad ideological consensus that existed among private citrus growers about the socioeconomic path to the establishment of the Jewish National Home.[4]

The person who expressed most emphatically and clearly the growers' views on this issue and many others is, without doubt, the writer, grower, and president of the Farmers' Federation, Moshe Smilansky. Below I attempt to assess Smilansky's views on this issue, under the assumption that his opinions about private capital and entrepreneurship in Jewish settlement of Palestine largely reflected those of his colleagues. Moshe Smilansky was a complex personality who deserves a biography of his own. A farm worker during the First Aliya years, a vintner and citriculturist, a leading player in various organizations of farmers, a delegate to two Zionist Congresses, president of the Farmers' Federation, a member of the Jewish Legion (Jewish units in the British army during World War I) and the Hagana, a loyal supporter of the Mandatory regime and the British Empire, and a prolific author of short stories about Arabs and Jews in Palestine, Smilansky shaped the consciousness of many members of the Zionist Movement. He was also an editor and expository writer who after only a short stay in Palestine had formed a clear opinion about the need for cultural and social rapprochement with the country's Arab population.

Moshe Smilansky was born in 1874 in a village near Kiev. His father was a leaseholder in a large estate and ran the farm of the estate for many years, developing a special affinity for this kind of vocation, especially where it concerned fields of grain. As an adolescent, as he attested in his multivolume autobiography, Smilansky was influenced by an amalgam of ideas. He came from a Hasidic family; his parents regularly sought the counsel of the heads of the Chernobyl dynasty. He was also influenced by the Hebrew literature and press of the time, the writers of Hibbat Tsiyyon (a late-nineteenth-century pre-Herzlian Zionist movement in Russia), private tutors who were "refugees" of the Bilu movement (a First Aliya organization that, in its ideology, preceded many concepts of the "mythological" Second Aliya), "progressive" ideas from the Mikhailovsky school, and Tolstoyan ideas of return to the soil and simple and moral life. Above all, however, Smilansky was a Hovev Tsiyyon (an adherent of Hibbat Tsiyyon), who, like many of his contemporaries in the movement, believed only a return to agriculture and "nature" could lead the Jewish people to spiritual and national renewal. Were this not enough, Smilansky was also a prominent follower of Ahad Ha'am (Asher Hirsch Ginsberg, 1856–1927), a philosopher and leader of "spiritual Zionism."

Smilansky landed in Palestine in 1891, settled in Rishon Lezion, and began his career as a laborer for Baron de Rothschild and various independent farmers. In the summer of that year, he was joined in Palestine by additional members of his family, including his brother Meir, his sister Shifra, and two nephews, David and Ze'ev Smilansky. In the ensuing years, the two last-mentioned also became well known personalities in the Yishuv. They were immigrants of means. Moshe Smilansky's father purchased six hundred dunams of land in the newly founded moshava of Hadera and deposited money for his family members with the fund of the executive committee of Hibbat Tsiyyon in Jaffa. After all the Smilanskys contracted potentially fatal cases of malaria (Moshe's sister, Shifra, never recovered from the illness and died from it several years later), they moved from Hadera to Rishon Lezion. There, Moshe Smilansky spent additional time as a laborer, but by 1893 he had acquired a parcel in Rehovot, an unaffiliated moshava, and lived there for the rest of his life.

In the first decade of the twentieth century, Smilansky was closely associated with Ha-po'el ha-Tsa'ir (a Jewish labor party founded in 1905) and frequently contributed to the movement's journal until he broke with the movement in 1911 due to what he said were disagreements on the "Hebrew labor" issue. Agile with a pen, Smilansky began to publish expository pieces, short stores, articles, and polemical writings in the Hebrew press.

Soon after he reached Palestine, Smilansky rose to prominence among the farmers and took part in most of their important organizations. In 1925, he established the Association of Farmers in Judea and Samaria (which in 1931 became the Palestine Farmers' Federation) and was its president and the editor of its journal, *Bustenai* ("The Plantationer"). After moving to Rehovot, Smilansky initially grew grapes, but as the profitability of viticulture declined and when Baron de Rothschild stopped subsidizing it, Smilansky was one of the first to uproot his vineyards and replace them with citrus orchards. In short order, Smilansky became a citriculture expert who attracted many followers and whose advice was widely sought. He derived his expertise not from formal schooling but from a combination of practical know-how, study of local cultivation methods, and gleaning of theoretical knowledge from pro-

FIGURE 2.1. Moshe Smilansky at fifty-three.

fessional literature. Smilansky began to assume a central role in the growers' organizations, and in 1930, shortly after he joined the veteran citrus marketing cooperative Pardess, became the chair of its executive committee.[5]

MODELS OF COLONIAL SETTLEMENT

In 1904, Smilansky published an article in the important journal *Ha-shiloah* titled "[Remarks] on Questions of Settlement in Palestine." The article set forth a comprehensive program for Zionist settlement of the country. Notably, at the time the Zionist Movement was still officially opposed to "practical work" in Palestine (although fissures in this policy had begun to appear, most conspicuously in the establishment of a branch of the Zionist bank, the Anglo-Palestine Company (APC; during the Mandate period it changed its name to the Anglo-Palestine Bank [APB], in Jaffa); and the Second Aliya immigrants had just begun to arrive. Smilansky's seeming intent was to use this forum, *Ha-shiloah*, to dispel the vagueness by presenting a comprehensive "Palestinocentric" solution.

The plan that Smilansky set forth was predicated on a specific settlement model, that of nineteenth-century European colonization. Basing himself on contemporary colonization literature that he had gathered (especially the writings of the French colonial settlement scholar and economist Paul Leroy-Beaulieu), Smilansky strongly differentiated between two motives for colonization: material gain and permanent settlement in a new country. The Zionist settlement enterprise, in Smilansky's opinion, was a venture of the latter type:

> The second kind includes colonization that aims to furnish people who have no remaining place in their old homeland, due to poor material conditions or national and religious persecution, with a new homeland. . . . They sought to naturalize themselves in a distant country, and we wish to do the same.[6]

Thus, in Smilansky's view, the motives of European Zionists and other Europeans who wished to find themselves a "new homeland" were one and the same. However, in his opinion, there were three perceptible differences between the groups. First, non-Jews were unconcerned about the identity of their country of destination, whereas Jews were certainly concerned. Smilansky was a man of Eretz Israel to his very soul, a stalwart of Tsiyyonei Tsiyyon—"Zionists of Zion," a leading faction in the early-twentieth-century Zionist Organization whose members believed, in contrast to Herzl and other Zionist leaders of the time, that Jewish national aspirations could take place only in Eretz Israel—and for him the Jewish "rebirth" *(tehiya)* could take place in Eretz Israel and nowhere else. The second principal difference between the Zionist settlement project and European colonialism concerned the component of "rebirth" in the process of Zionist migration to Palestine. The essence of rebirth, he maintained, is return to and cultivation of the soil. Only by cultivating the soil of Eretz Israel, he said, could the Jews redevelop their "love of homeland." However, Smilansky considered agriculture only the beginning, not the end, of the rebirth process. After the Jews regained, by means of agriculture, "the naturality of which the unnatural state of exile has deprived them," they would be able to develop industry, trade, and culture in Palestine as well. The final goal, he stated, is "the redemption of [the People of] Israel on the soil of Eretz [the Land of] Israel," by means of acquisition of the Land; mass settlement and the establishment of self-rule, "like any free people that dwells on its soil." This, however, Smilansky admitted, was a distant goal. Until it could be attained and in order to attain it, an explicit plan for the present should be elaborated.

It was in respect to the practical process of settlement that Smilansky found the third difference between European colonists who settled in some "unsettled part of the world" and Jewish colonists in Palestine. "Our country is not an unsettled wasteland," Smilansky stated, reinforcing one of the principal truths that Ahad Ha'am had proclaimed in his first *Emet me-eretz yisrael* (Truth from Palestine) article, published in 1891. The "unsettled" lands where Europeans wished to go were inhabited by "savages or semi-savages" who could be bought off or killed; afterward, the new settlers could simply seize their land. In contrast, "The inhabitants of our land are not savages who may be treated with disregard." Accordingly, the acquisition of land in Palestine must not be forcible. "We are coming to redeem our land," Smilansky maintained, "not to grab it." These remarks express the crux of Smilansky's settlement philosophy: to treat the Arabs of Palestine considerately, in both humanistic and national terms, without waiving the final goal as he envisioned it, that is, acquiring most of Palestine and settling it with a Jewish majority.

Smilansky's copious writings and articles as a citrus grower and a leader of the farmers (and his series of short stories, *B'nei 'arav* [Sons of Arabia]), reveal another dimension in his attitude toward the Arab inhabitants of Palestine—an "Orientalist" one. On the one hand, Smilansky evinces a romantic appreciation, with Tolstoyan undertones, of the connection between the Arab peasant *(fellah)* and citrus grower with the land and nature; he urges the inexperienced Jewish farmer to learn from the "son of Arabia," who is close to nature and the soil. On the other hand, he speaks in a disdainful and condescending tone, that of a scion of culturally and technologically "advanced" Western civilization, to the selfsame peasant, who turns out to be "primitive," "savage," and "slavish."[7]

Smilansky adopted the European "settler society" colonization model, reduced it to several main components, and pondered the extent to which these components should be adjusted to the conditions of the Zionist Movement and Palestine. The components were (1) a settlement charter; (2) purchase of land; (3) settlement method and vocational patterns; (4) the type of settler; and (5) relations between the colonists and the "metropolis," that is, the Jewish Diaspora.

1. Smilansky defined the *charter* that the Zionist Movement sought from the Ottoman authorities as identical to the franchise that various European countries granted to settlement companies.[8] Smilansky did not consider it wholly inconceivable that the Zionist Movement might eventually receive a charter for Palestine. However, he asserted, it was neither reasonable nor desirable to obtain such a charter at the present time. It was unreasonable because the Turkish authorities would not grant it and because Palestine was inhabited by people who owned most of its land. A charter alone would not transfer title over the land to the Zionist Movement. It was undesirable because the Zionist Movement could not assume

so heavy a financial burden and because political rights did not confer title to land. Land could not be brought under national control unless it were first purchased and cultivated.[9] Accordingly, Smilansky advised against the idea of seeking an inclusive charter and argued in favor of a policy of *incremental settlement*.

2. Land purchase, according to Smilansky, is the main condition for the attainment of the Zionist Movement's ultimate goal. In the act of "redeeming the land" (the biblically inspired Zionist term for land acquisition in Palestine), not only the institutions of the Zionist Movement but also private capital and initiative were of central importance. This was another insight that Smilansky acquired from his study of nineteenth-century European colonization, in which private enterprise and capital had been integral (in either a leading or a complementary role) in the settlement act. Evidently, Smilansky preferred to see the Jewish National Fund (JNF) handle all land acquisition but admitted that private initiative would have to be mobilized as well due to the Zionist Movement's penury. Thus, Smilansky proposed *a division of labor*: the JNF would purchase large parcels of land on which large-scale preparation, infrastructure work, and settlement would be possible, while private developers would acquire small parcels and thereby contribute to the national effort of "redeeming the land." In this manner, too, the private developers would not forfeit "personal utility," their primary aim, since they would be able to resell their land or settle and cultivate it themselves. Here we see another important fundamental in Smilansky's settlement philosophy, which he continued to espouse during the Mandate era: the principle of a division of functions between private and national capital and rejection of the exclusivity of either.
3. The correct method of settlement and economic endeavor, in his opinion, were embodied in one word: *agriculture*. It was conventional wisdom in some Zionist circles that urban forms of settlement, that is, industry and trade, were the fastest and the economically and socially safest ways to settle Palestine. The most conspicuous champion of this outlook was the director of the Anglo-Palestine Company, Zalman David Levontin.[10] Smilansky dismissed this view on ideological grounds. The Jewish rebirth, he argued, should be instigated by agriculture as a prerequisite for all other economic activities. The European colonization experience had shown this to be so, he added. Thus, agriculture should be given priority in the initial phases of the Zionist settlement endeavor. The results, Smilansky concluded, would be twofold: social and national rebirth and control and development of the land. Only later would the other phases follow, resulting in the incremental and "natural" creation of a new Jewish society composed of people of all kinds, all classes, and all economic pursuits.

4. If Smilansky tended to believe that land acquisition would be spearheaded by national capital, in respect to the type of settler he clearly preferred persons of *middle-class affluence* and of what he considered the appropriate emigration age, the twenty to forty bracket. According to Smilansky's program, most settlement work would be done by private "colonization companies" that would operate in coordination with the "metropolis" (the Diaspora). These companies would buy the land, invest in its infrastructure (roads, irrigation systems, etc.), purchase livestock and physical plant for the settlers, and sell both the improved land and the live and physical inventory to the settlers. Thus, in Smilansky's vision, Zionist settlement would be mainly a private agricultural venture financed with private capital, undertaken at private initiative, and situated on private land. Private settlement, he believed, had another advantage: by its very definition, it diminished the use of philanthropy and would stanch the incursion of undesirable elements, as he viewed them, that had socialist or religious intentions.
5. As for relations between the colonies and the "metropolis," Smilansky cited his colonization theory to sketch a relationship that would involve subvention without custodianship. Smilansky, like his mentor Ahad Ha'am, categorically rejected all forms of custodianship. In his way of thinking, management of a settlement venture by white-collar types from the "metropolis," stationed in the colony, would lead to enslaved farmers, moral corruption, and economic inefficiency. Only farmers who bore responsibility for the results of their actions would practice economic efficiency and place their farms on a course leading to profitability. Therefore, Smilansky proposed—again basing himself partly on the European colonial experience—that the moshava be given operational freedom without forfeiting its entitlement to material assistance from the "metropolis."

The main elements of Smilansky's construct—a visible Zionist national consciousness, affirmation of agriculture as a value and a path to economic viability, the centrality of private capital and initiative in building the new Yishuv in Palestine, assertion of individualism and personal freedom as general human and basic values in the praxis of private initiative, negation of any hint of philanthropy, and affirmation of the use of public and national capital—recurred with different degrees of emphasis in Smilansky's writings not only during the Ottoman era but also during the Mandate years. However, Smilansky wrote the article at issue here with the clear sense that he possessed the power and knowledge to set forth an inclusive framework of large-scale action, and the article intended to do just that. Only due to the situation of the Zionist Movement and the Yishuv at the time could Smilansky speak in such a tone.

Matters developed differently. Within a few years, on the eve of World War I, and with greater intensity during the British Mandate era, Smilansky's articles and writings—like those of other citrus growers and private entrepreneurs—spoke in a *defensive* tenor. The defense expressed in these writings was often very vigorous if not aggressive, but Smilansky and his comrades had clearly lost the position of primacy, of people at the cutting edge of action.

In the decade between Smilansky's article and the beginning of World War I, the Yishuv underwent several important changes that had a perceptible effect on the social and ideological status of private capital and enterprise in Zionist history. On the one hand, the private sector flourished and expanded: The First Aliya moshavot, foremost those in the southern part of the country ("Judea," as it was called at the time), became prosperous, grew socially and demographically, and expanded their land area. The nonagricultural segment of the private sector, which accounted for most of the Yishuv, also expanded during this time. First indications of industry began to appear. The founding of Tel Aviv in 1909 marked the inauguration of this all-Jewish urban center, which grew at an impressive pace. On the other hand, the Zionist administration in Palestine developed and began to cooperate closely with a new element in the Yishuv, the working class. The political culture and the social and cultural structure of the Yishuv underwent a radical change when the Second Aliya immigrants arrived and organized ideologically and politically. In 1908, about four years after the onset of the Second Aliya, the Zionist administration cemented its place in the Yishuv by establishing the Palestine Office, under Arthur Ruppin, to centralize the practical settlement activity of the Zionist Movement.

The Second Aliya immigrants acted with considerable alacrity to fashion a complete ideology that ruled out the private-settlement outlook as the moshavot practiced it. Their motives were both socialist, pertaining to the "exploitation" of the worker, and national: private farmers, needing large quantities of seasonal labor to cultivate their vineyards and orchards, employed not only Jewish workers but also (and mainly) Arab workers. The leading spokespersons of the Zionist Labor Movement in Palestine portrayed Arab labor in the farming moshavot in terms culled from the most hallowed domains of Jewish tradition—"desecration of God's name," "defilement of the sanctuary," and "idolatry"—and accused private farmers of violating the most sacred values of Zionism in its Labor version. Private farmers were devoid of, or at least weak in, national consciousness. They were typical European "*kolonistim*" whose sole motivation was the prospect of immediate financial gain.[11] As Ben-Gurion expressed it, "The private farmer is not only unmoved by the whole cause of social justice and equality but is also unconcerned about the national future of the Yishuv—whenever it might cost him something."[12]

Concurrently, Ruppin began to take a disapproving stance toward the moshavot and everything that, in his opinion, they represented. Ruppin, as Derek Penslar has shown, had held socialist and vitalist-national views even before he had come to Palestine, regarded the working class as the crucial agent of the Jewish revitalization. The workers, he believed, would revitalize the Jewish nation and awaken it to rebirth. Like Smilansky and many of his contemporaries in the Zionist Movement, Ruppin regarded return to the soil as the prime condition for the national transformation of the Jewish way of life. Accordingly, even though Ruppin affirmed, in principle, the importance of private capital in building the Yishuv he cooperated mainly with the workers as the years passed and was increasingly partial to their views on private capital and its role in agricultural settlement. While Ruppin did not rule out cooperation with private capital, he took clear exception to the involvement of such capital in settlement ventures that were undertaken at the initiative of the Palestine Office.[13]

In a summarizing lecture on "Our Settlement Method," delivered at the XI Zionist Congress (1913), Ruppin explained why he refused to co-opt the private-enterprise moshavot in his settlement enterprises. The moshavot, he said, were an "aging organism" that could be restored to "youthful vigor" only by "introducing new enthusiastic young elements from Europe." The moshava method of settlement might be profitable, Ruppin admitted, but the farmers would not cultivate the land with their own hands and, what is more, most wage labor that they would employ would certainly not be Jewish.

Ruppin's critics noted that the settlement locations sponsored by the Palestine Office were incurring steady losses. Ruppin countered them by claiming that it was not his goal to prove that settlement in Palestine could become profitable; after all, the moshavot had demonstrated this in their business activities. From his standpoint, the national revival entailed not only a return to the soil but also a socialist vision of equality and social justice. Therefore, Ruppin argued that the training of workers on the Zionist farms or "restoring one moshava to new life" was worth the financial losses. On this basis, Ruppin drew a dichotomous portrayal of private initiative and national endeavor. In his opinion, one might argue "with absolute certainty that any enterprise in Palestine that generates a large profit for the merchant is usually the least profitable of all for our national goal, and vice versa."[14]

Thus, these two central sources, the Zionist Labor Movement and important personalities in the Zionist Movement, joined at this formative time to create a linkage, perceived as negative, between private capital and enterprise and the engineering of a Jewish national rebirth.

THE "BOAZES"

In the spring of 1912, with these matters in the background, Ahad Ha'am published his article "All in All" *(Sakh ha-kol)* in which the progenitor of the "spiritual center" idea summarized the impressions of his most recent visit to Palestine.[15] In his article, Ahad Ha'am coined one of the most powerful similes that would ever be applied to the private moshavot, a turn of phrase that would remain vital in the Mandate era as well. In contrast to his earlier writings, "Truths from Palestine," the Zionist leader returned from his latest visit with a clear inner sense that his vision of the establishment of a "spiritual national center" in Palestine was marching toward fruition. He admitted that those doing the work still proclaimed the old truths, that their sole purpose was to create a "safe haven" for Diaspora Jews and not a "spiritual center." However, their "sense of national existence," of which they were unaware, was guiding them to act in contravention of their rhetoric and to move toward the fulfillment of his vision.[16] What was happening in Palestine, Ahad Ha'am claimed, was not the fulfillment of the Herzlian Political Zionism; after all, there was no charter from the Ottoman authorities and Jews were not flowing to Palestine en masse. The developments in Tel Aviv, Jerusalem, the Hebrew schools, and the moshavot were in fact the establishment of the "spiritual center."

Obviously, Ahad Ha'am had much to say about doings on the moshavot. "Traditional" Zionism envisioned the evolution of a class of rooted Hebrew farmers in Palestine. However, the formation of such a class had not occurred, the leader of "spiritual Zionism" claimed, and could not possibly come to pass.

> Even if our rural Yishuv in Palestine eventually expands to its widest possible limit, it will always remain an elite community of a developed cultural minority whose strength lies in its minds and capital. However, the large rural masses, whose strength lies in their manual labor, will not be ours even then.[17]

It is true that the moshava farmer heads for his orchard or vineyard in the morning, is not averse to working the land with his own hands, and is closer to the land that any urban Jew. However, "he is not totally absorbed in the land" and wishes to lead "a genteel life, to enjoy the fruits of his contemporary culture, the pleasures of the body and the soul." Therefore, by his very nature and his social and cultural essence, he must augment the sweat of his own brow with wage labor.

Notably, in his "Truth from Palestine" Ahad Ha'am had in fact advocated the creation of a class of "real" farmers who would form the basis of the "spiritual center." The establishment of such a foundation was crucial in his perception of the spiritual center. There, in the embryonic Hebrew center in the Land of the Patriarchs, "the Jew will also elevate himself as a member of humankind. He will wrest his bread from the soil by the sweat of his brow and will create living conditions for and by himself in the spirit of his nation." Clearly adopting the criticism of the Jewish concentration in "unproductive" industries, Ahad Ha'am advocated the "expansion of colonization" on the basis of "agriculture, not 'agribusiness' and, perforce, by trained farmers and not by calculating merchants."[18]

Two decades later, however, when he published "All in All," the Zionist leader recanted his earlier views. In a feat of rhetorical sleight-of-hand, Ahad

Ha'am dissociated himself from the idea of developing a class of "real farmers" in Palestine and attributed this idea mainly to the Political Zionists, the "safe haven" people, as he termed them. Consequently, in "All in All" Ahad Ha'am described the moshava enterprise in glowing and strongly positive terms even though it had not generated a class of "natural" farmers and had based its economy on everything that he had decried two decades earlier—"agribusiness" and wage labor.

Accordingly, Ahad Ha'am termed the moshava farmer "akin to [the Biblical] Boaz in his time." To Ahad Ha'am's mind, this analogy, based on the Hebrew farmsteader who bonded with his land and field but did not personally cultivate it, leaving this task to wage laborers, was altogether favorable. However, members and supporters of the Labor Movement construed it as the epitome of everything that was negative about the private agricultural settlement enterprise. Boaz's "boys," as the Biblical narrative terms the hired farm workers, now evolved into Arabs. From then on, the farmers of the old moshavot were repeatedly portrayed as "Boazes" who lived off strangers' labor in flagrant disregard of the Jewish national singularity.

By 1913, a year after "All in All" saw print, Smilansky had to refute a new accusation against the old moshava farmers: "Boazism." In a pungent article in the journal *Ha-po'el ha-tsa'ir* the agronomist Yitzhak Wilkansky, manager of the Zionist farms in Ben Shemen and Hulda, attacked Jewish private enterprise in Palestine and accused it of being antinational. In his article, replete with lyrical expressions culled from the trove of traditional Jewish culture, Wilkansky accused both the moshavot and Tel Aviv of developing a colonialist community that had only profit in mind and could not bring redemption to the Jewish people. Reliance on private capital, coupled with the profit motive, was leading to the employment of cheap Arab labor and, by necessity, to dissociation from Eretz Israel, the soil, and the entire national rebirth process. The Yishuv derived its national life neither from the moshavot nor from "the bourse in Liverpool" (where Palestine citrus was sold), said Wilkansky in his critique of the "Boazes," but from the "Hebrew worker" and "our national institutions." Only due to the "heart" (the living core) and the "hands" of the Hebrew worker could "private capital" grow and flourish, and only with "the nation's money" and not that of "private enterprisers" could the Jewish entity in the new Yishuv attain vitality. The private farmers, Wilkansky claimed, had contributed nothing to the growth of the Yishuv; everything they had originated "in the Baron's money and power and the muscles of the Ishmaelites [Arabs]."[19] The dichotomy was clear: Private capitalists were antinational and antisocial; only "national capital" carried the torch of the Jewish national and social rebirth.

Although every line of Smilansky's response to these allegations bespoke defensiveness, his arguments sought not to justify but to set forth a complete worldview, anchored in evidence and statistics, that affirmed the role of private capital and initiative in building the Jewish national entity in Palestine. As for the national element in his thinking, Smilansky did not discuss this aspect at all; his remarks show that he considered it self-evident. His very presence in Palestine and his participation in the nascent Yishuv said as much. As for the economic strategy that should to be taken in building the Yishuv, Smilansky probed it from an inclusive perspective: the extent of its components' contribution to the attainment of the final goal, the purchase of land in Palestine and the creation of a Jewish majority there. To achieve these aims, the prime requisite was the creation of a true economic base—as opposed to an affluence based on subsidies—that would abet the expansion of the Jewish-controlled land area and create reliable sources of livelihood for as many Jews as possible. Therefore, the key word in his article is *philanthropy*. Wherever private initiative was given free rein, Smilansky claimed, the Yishuv flourished economically and spiritually. Wherever inhabitants were not allowed to act at their discretion and wherever a mechanism and a climate of philanthropy prevailed, economic losses and torpor were evident. Only an individual who fought for his sustenance, Smilansky argued vehemently, acts with economic efficiency. What is more, only such a person seeks new entrepreneurial opportunities in the fields of business or technology.

> Only the *individual*, fighting for his existence, only a private beginning and private capital, can survive under the harsh conditions that beset our settlement effort. Only private initiative can repel [the difficulties], only it can bear fruit, only it can find experienced people to whom it may entrust its fate, and only it can create *healthy* conditions of labor and endeavor. Beyond the limits of private initiative lie death and stagnation.[20]

Due to their wish to maximize their profits, the moshava farmers had improved the orchard and citrus industries, claimed Smilansky. Their orchards now became productive five years after planting, whereas the Arabs' orchards did not yield fruit until the eighth year. Furthermore, the introduction of superior growing methods had increased the yield per dunam. Much the same had happened in viticulture: Since this industry had gone over to private farmers, it had been producing "dividends" instead of deficits. Smilansky credited the business success of the Anglo-Palestine Company largely to "profits that [APC] earns *from us* and the deposits that *we* have placed with it" (emphasis in the original; in several articles, Smilansky also expressed strong appreciation of Levontin's management of the bank). Summing up, Smilansky stated that among the moshavot in Judea those based on private capital and initiative had already achieved financial self-sufficiency and had become the economic "arteries" of the Yishuv. Smilansky did not overlook Tel Aviv, stressing that this flourishing urban center was being built by private initiative only. Admittedly, the JNF had given the founders of Tel Aviv an initial loan to establish their neighborhood. (Smilansky had opposed this transaction in the belief that the JNF should focus on land acquisition.) Now, however, such a sum could be raised on "one streetcorner" of the new city. Countering the claim that the capitalist economy was prospering only by dint of Jewish workers' labor, Smilansky noted that "hundreds of workers in Hadera, Petah Tikva, Rishon le-Tsiyyon, Wadi Hanin, and Rehovot" were able to support themselves solely due to private capital.

Notably, this article and Smilansky's subsequent writings give evidence of a slight change in Smilansky's attitude toward agriculture in the Jewish rebirth effort. The earlier categorical demand to entrust all cultivation to repatriated Jews yielded to a more general discussion of Jewish agriculture. In his memoirs, Smilansky repeatedly spoke sadly about having had to stop working the soil almost totally due to his age, his many public duties, and the expansion of his farmholdings. He even had to transfer the management of his lands to hired managers, he admitted ruefully. However, in both his memoirs and his articles he always expressed his belief in the existence of a Jewish agriculture sector as an essential condition for the Jewish rebirth. Accordingly, here and in additional articles, Smilansky pointedly stressed the importance of a strong Jewish agriculture industry that could sustain a "natural" Jewish working class. Clearly, then, the reality of prosperous private Jewish farms based on wage labor such as Smilansky's had helped to change the outlook of Smilansky the farmer and writer.

Smilansky also disputed the notion that the "conquest of labor" was tantamount to crowding out Arab workers from the Jewish economy. He saw no contradiction between his national consciousness and the employment of Arabs in his vineyards and citrus orchards. In many articles at the time he stated repeatedly, with emphasis, that while an effort should be made to place most labor in Jewish agriculture in Jewish hands, any attempt to oust the Arabs from the Jewish economy should be categorically opposed.[21]

Notwithstanding the success of the private Jewish farm, Smilansky did not change his mind about the role of public and national capital in the settlement process. Here, too, he believed it necessary in the Zionist enterprise—as in any other colonizing endeavor—to use public and national capital for the investment in infrastructure. This was the "cornerstone" that Baron de Rothschild had laid for the Yishuv; indeed, in Smilansky's opinion, Rothschild had performed the service that a European government would perform for its colonization projects. Now that the cornerstone had been laid, however, "free capital" should be used for the construction phase. The British colonies had been successful because Britain had turned over the building effort to private capital. The French, Portuguese, and Dutch colonies had failed, Smilansky continued, because these countries had not handed their management to private enterprise. Only recently had economic life in Algeria changed for the better because "private capital had been attracted" to that country.[22] Rothschild had been mistaken on that score, since only after he withdrew his patronage from the moshavot did they begin to flourish economically and spiritually. In contrast, moshavot that were still under custodian administration, such as those of the ICA in the Galilee, evinced no growth whatsoever and were actually declining.

Smilansky then used the same criteria to examine the Palestine Office's settlements. These settlements, he opined, were blatant examples of the illegitimacy of the custodianship method. Ruppin considered them the foundation that would bring life to the new Yishuv, whereas Smilansky saw in them nothing but "death and stagnation." "Kinneret began with great commotion and went bankrupt furtively," Smilansky sniped at his critics. "In its four years of existence, it created nothing but enormous financial losses." The farms in Ben Shemen and Hulda also performed miserably, recording nothing but losses, and although he found encouraging indications at Merhavia, the losses incurred by this cooperative venture—some 60,000 francs in a two-year period—gave him reason to believe that this attempt, too, would fail for reasons of "philanthropy." 'Ein Ganim, the workers' cooperative near Petah Tikva, was not an example of a successful settlement financed with national capital, Smilansky claimed, because its entire existence depended on "private capital in Petah Tikva," where the inhabitants of 'Ein Ganim made their living. Smilansky did find much to commend in Degania (the first Kibbutz), with its autonomous administration. However,

> Degania will neither secure Eretz Israel for the Jews nor populate the Yishuv by means of moshavim such as 'Ein Ganim that will grow alongside it; instead, it is the *capitalistic Petah Tikva* that will accomplish this. And if you say that one could create any number of Deganias with the same amount of capital as resides in Petah Tikva—that is the whole problem, gentlemen! Petah Tikva can attract the capital and Degania cannot.[23]

Land acquisition in Eretz Israel, Smilansky argued repeatedly, depends on private capital and the initiative to use it. Only private capital, acting for the profit motive, can drive business entrepreneurship in agriculture, trade, and industry that will result in economic efficiency and growth. Only thus will the Jewish-held area grow and will large numbers of settlers be able to make a living. For this reason, Smilansky also opposed the principle of nationalization of land, viewing it as an obstacle to the optimum selection of settlers and the economic development of agriculture. When public and national capital does not content itself with preparing conditions for settlement and engages in managing the settlement effort outright, it actually becomes philanthropic capital. The results—in the ICA settlements and in the farms run by the Palestine Office—are losses and scanty population.[24]

The onset of the British Mandate for Palestine in July 1920 was a momentous change from the standpoint of the Zionist Movement and the

Yishuv. A legal framework for large-scale Jewish immigration and convenient purchase of land had come into being.[25] At this time, the Zionist Movement cemented its principles as to methods of settlement and immigration. The resolutions of the Zionist Conference in London (July 1920), which drew the contours of the settlement policy of the Zionist Movement, did leave some room for private capital in land acquisition and did not exclude financial aid for private settlement ventures. However, it was the general trend of thought to invest resources mainly in settling "national" land with settlers who accepted the principles of the Jewish National Fund and, especially, the principle of Hebrew labor. The conference also expressed the principle of "self-labor" (labor by the farmer-settler only) in "national" settlement as a categorical imperative and resolved to cooperate closely with "workers' institutions" in all phases of the settlement endeavor, from immigration to possession of the land and education.

One of the main underlying rationales in these decisions, as Jacob Metzer has shown, was concern that private capital, operating under imperfect capital market conditions, might create latifundia (large estates of the East European type) that would monopolize the factor-inputs market. Were this to occur, private farmers would be able to lower farm workers' wages and eventually bring in cheap Arab labor.[26] Labor Zionism, as noted, considered this a flagrant manifestation of nonnational thinking. Ben-Gurion expressed this view in the following way: "A *hovev tsiyyon* [lover of Zion] such as Ahad Ha'am, who contented himself with Boazes who lived off strangers' labor and neither required nor believed in Hebrew labor, did not believe in Zionism."[27]

To forestall the "insidious" effects of capitalistic economics, the Zionist Movement, by means of the labor organizations, adopted a settlement policy that aimed to offer an antithesis to the citriculture economy. In the Jezreel and Jordan valleys, where the Labor settlement enterprise was centered in the early years of the Mandate era, "mixed farms" were established for the purpose of using "self-labor" to achieve nondependency on the market.

Berl Katznelson summed up the matter nicely: "The aspiration to self-sufficiency" (i.e., economic autarchy) was "a rebellion against the citrus-grower outlook that was dominant in Palestine at the beginning [of the Second Aliya]."[28]

THE LABOR MOVEMENT CHANGES ITS ATTITUDE TOWARD PRIVATE CAPITAL AND CITRICULTURE

The Fourth Aliya (1924–1928) spurred the capitalistic tendencies in Jewish economic activity. It is true that these tendencies initially led to economic and demographic woes that Zionist historiography calls the "Fourth Aliya crisis." However, the great momentum in citriculture—for the Jewish and Arab sectors alike—began in 1926, before the crisis ended and, as stated above, pushed the Yishuv's economy more toward a capitalistic type of economic life. In view of these developments, the Labor Movement abandoned the idea of achieving economic consolidation mainly on the basis of self-sufficiency. Notably, by the 1920s the Labor Movement had already moved into pronouncedly capitalistic fields of activity, for example, the establishment of Bank Hapoalim and the operations of its construction company Solel Boneh.[29]

The Fourth Aliya, as stated, pulled the Jewish economy in a capitalistic direction. The change was evident in the development of Tel Aviv but evinced itself in the vigorous growth of citriculture as well. Organized groups of middle-class private capitalists began to settle in the Sharon area, where little Jewish settlement had occurred to that time. After several years of financial hardship, quite a few of these groups—including Benei Binyamin, an organization of farmers' offspring—asked the Rural Settlement Department of the Zionist Organization for financial aid. The department, which until then had not engaged in private settlement activity, lent money to the new moshavot and developed programs with which they could establish "mixed farms," including, for the first time, citriculture.[30]

What is more, the Labor Movement agreed to cooperate with private capital even in its own bailiwick, agricultural settlement. Again it was Berl Katznelson (in 1933) who aptly summed up this turnaround in the attitude toward private capital:

FIGURE 2.2. David Ben-Gurion speaks before Jewish workers in Jerusalem, 1924.

> When we began, before we grew in numbers and strength, we were afraid of private capital that had nothing but profit in mind. Twenty-five years later, we attained a level that allowed us to take into account working [labor] settlement with private capital as well, either as a creator of jobs or as capital seeking an investment and willing to consider even the worker as a suitable vehicle.[31]

Views expressed in the early 1920s against the "self-labor" principle resurfaced now. For example, Arthur Ruppin argued "heretically," not for the first time, that the "self-labor" idea was hindering the development of the Yishuv by preventing the employment of Jews in the Zionist public farming settlements. It was acknowledged that even medium-sized mixed farms needed more unskilled labor than the Labor settler and his family could provide. Consequently, there were occasional de facto violations of the "self-labor" principle. In some cases, contract groups were hired at ploughing or harvest time; in other instances, Labor settlers continued to hold jobs off the farm and hired day laborers to cultivate their orchards in their stead.[32]

Apparently, however, the changeover to cooperation with private capital represented a broader turnaround in the economic activities of the Labor Movement. One may characterize it as the adoption of capitalistic methods of thought and action by the economic organizations of the Histadrut. Although the issue undoubtedly deserves thorough examination, research has occasionally discussed various aspects of the new capitalistic leanings in the Labor movement's thinking and economic activities. The studies and various contemporary sources show that, in view of the failures of Hamashbir Hamerkazi (the Histadrut's central marketing cooperative) and Solel Boneh in the Fourth Aliya crisis, the Labor Movement underwent a change in the direction of greater prudence in its business activities.

The Fifth Convention (in late 1926) of the leading political party of the Labor Movement, Ahdut ha-'Avoda, was supposed to discuss the results of the Fourth Aliya crisis, among other things. In advance of the convention, Ben-Gurion and Katznelson (the party's leaders) drew up a comprehensive philosophical statement about the workings of the Labor economy in the period following the Fourth Aliya. Labor's goal of establishing, from the ground up, a Zionist society that would also be "socialist" had proved "utopian," as Ben-Gurion put it. No "socialist economy," that is, an economy in which means of production are nationalized and class differences are eradicated, had been established in Palestine. Instead, a "labor economy" had come into being. In its outward manifestations, Ben-Gurion asserted, a labor economy was a capitalistic one in every sense. Ben-Gurion admitted that in the internal economy of kibbutz 'Ein Harod, for example, "communistic relations are dominant." However, "Relations between 'Ein Harod and other settlements—private settlements and Labor Movement settlements alike—are ordinary relations, like those in any capitalistic economy." Therefore, in the aftermath of the Fourth Aliya, he observed that

> [t]he Hebrew economy under construction in Palestine rests on capitalistic foundations. A Hebrew class society is coming into being in Palestine, and even the Labor economy that is taking shape in Palestine is taking on a capitalistic form in its outward actions.[33]

Katznelson addressed the Ahdut ha-'Avoda convention in a similar vein. The Fourth Aliya, Katznelson argued, had transformed the economic reality in Palestine. One could no longer speak of "negation of the market." The opposite approach was needed—an attempt to take over the market by invoking its own weapons, namely, capitalistic methods. Thus, "The way to liberate [ourselves] from market exploitation" is, paradoxically, by "developing industries, studying the market, and going into systematic commerce." Candidly, Katznelson explained that the Labor agricultural economy, based on "self-sufficient" types of activity, could hardly cover its labor expenses, let alone other outlays such as investment payback and amortization. In regard to Hebrew agriculture at large, however, Katznelson observed that "the activities that have made breakthroughs are profiting not due to field crops and self-sufficiency but due to *export activities*" (emphasis added). By implication, Katznelson concluded, summing up the contours of the new policy, every Labor's economic unit "must strive to maximize its market utility and 'profit'."[34]

A comprehensive assessment of the implementation of this plan would fall outside the thematic limits of this book. Apparently, however, the operations of entities such as Tnuva and Tnuva Export (marketing cooperatives of the Labor Movement), which operated in the free market and did so on the basis of customary free-market patterns, corresponded to the trend described above. Furthermore, the Labor Movement established its own citrus planting cooperative, Yakhin. Yakhin's entrepreneurial activity as a Histadrut-owned enterprise—which made it a player in the "public sector" of the Yishuv—in the private plantation industry is described later in this book. Our concern here is the ideological change that the Labor Movement underwent in its attitude toward private capital.

Thus, even in citriculture, which the Zionist Movement had perceived to that time as the embodiment of all the "illnesses" of the capitalistic economy, the situation was changing. In locations where the "conquest" of Hebrew labor had been fully successful, such as private moshavot such as Magdiel and Ra'ananna, references to private capital were highly favorable. Even Ben-Gurion concurred:

> Private capital serves the Zionist cause if it functions as an instrument for Hebrew labor, national capital disserves Zionism if it does not abet this goal.

> The national capital [that Baron de Rothschild] invested in the construction of Zikhron Ya'akov and Binyamina missed its Zionist target. The private capital that built Magdiel and Ra'ananna performed a Zionist mission and did so perfectly.[35]

The total identification of private capital as an agent of capitalist "exploitation" and "antinational" Arab labor, as propagated in the formative Second Aliya era and accepted axiomatically in the early Mandate years, underwent a change: Private capital was no longer perceived as the epitome of evil. However, Ben-Gurion drew a firm distinction between ownership of capital and its purpose. "It is the *use* of capital, not its *ownership*, that determines its national value," he explained. Private capital can serve national needs, that is, economic activity without Arab labor. In such a case, it should be considered benign. However, if it serves a private purpose, it should be treated with suspicion and placed under national restrictions. By the same token, public capital, such as that of Baron de Rothschild, the ICA and PICA companies (which Ben-Gurion, from the perspective of 1932, considered part of the "national" capital), when used to further private enterprise and to employ Arabs in the Jewish economy, is contemptible and useless for the attainment of national goals.[36]

In his famous 1925 article, "The National Goal of the Working Class," Ben-Gurion ruled that only the working class could bring about the attainment of national goals because only its needs were "objectively" identical to the nation's. The interests of other classes inevitably clash with those of the nation. Since a private investor cannot but employ Arab laborers on his farm, his class interest surmounts the national interest.

Ben-Gurion's 1932 article, which legitimized the private citriculture settlements of Ra'ananna and Magdiel, did not swerve from this basic approach. From his standpoint, the capitalist system itself necessitates the use of non-Jewish labor. "A private investor cannot and will not make his wealth available for Zionist fulfillment and for it alone—because this clashes with his basic nature and his class inclination." Ben-Gurion acknowledged the possibility of individual exceptions but argued here, as in his 1925 article, that the Hebrew capitalist and the Hebrew worker are separated by an immutable objective difference in regard to the national interest. Thus, despite their shift from outright rejection of private capital, the leaders of the Labor Movement continued to frown on the private economy as an economic system.[37] Amidst this debate, the "Boaz" depiction of citrus growers always remained in the background and came to the fore whenever necessary.

"NATIONAL" SETTLEMENT AND "BOAZITE" SETTLEMENT

The "Boaz" simile occurred, for example, in remarks by Chaim Weizmann at a conference of British Zionists in 1933. Weizmann, in between his two tenures as president of the Zionist Organization, devoted most of his speech to economic affairs and lauded both the Yishuv's economic situation and the state of Jewish immigration. Weizmann even praised, "for the first time in twelve years," as one of his supporters said, the role of private capital in building the Yishuv and settling Palestine. Naturally, Weizmann's respectful comments pertained not to private enterprise at large but mainly to private enterprise in the settlement endeavor. However, Weizmann, like Ben-Gurion before him, attached a condition to his favorable attitude toward private settlement—"that the settler use Hebrew labor." Weizmann also established two categories of "nationalness" in various types of settlement. There are two kinds of settlers, he said: the *haluts* (pioneer) and the Boaz. It is the haluts, Weizmann explained, who prepares the land on which the timorous Boaz, who by nature "goes only where he's somewhat sure of his life and property," settles. Furthermore, the haluts settles more people per unit of land. The Boaz buys a hundred dunams and settles only one family there, his own. The haluts, who settles with financial assistance from the national funds—the role and necessity of which were now being challenged due to the success of the private farm, and which Weizmann now came forth to defend—managed to settle ten families on the same hundred dunams. Thus, Weizmann concluded, only halutsic settlement is "national settlement"; "Boazite" settlement is worth much less in national terms. The "Boazite" analogy recurred in the context of private settlement, which, Weizmann charged, lacked firm national conviction and had interests that clashed with the national interest in terms of settlement locations, the settler population, and the tendency to employ Arabs.

In response to Weizmann's remarks, Moshe Smilansky explained his philosophy on Jewish settlement in Palestine in two articles in the farmers' journal *Bustenai*.[38] He began by remonstrating vehemently against Weizmann's dichotomy of "'national' and 'Boazite' [private] settlement, which our leader seems to regard as non-national." Smilansky's main contention had not changed since the end of the Ottoman era. He stressed now, as he had then, that his national goals were no different from Weizmann's and that "the leader has got to be more cautious, responsible, and judicious in [his use of] the term 'nationalism.'" Smilansky's nationalism, the writer explained, was manifested in placing more territory in Jewish hands, expanding the Yishuv, and enhancing its economic growth. Private investors had outperformed the Zionist Movement institutions in land purchases, Smilansky noted. To marshal capital for land purchase, Smilansky argued, moral exhortations were not needed; one needed only to change the Jewish Agency's land policy and allow people to buy private land. Until the policy changed, he warned, people would refuse to contribute to the national funds.

Smilansky answered the question of the Yishuv's population in the same way. Repeating his argument from the Ottoman era, he stated that they would come only if they had a source of livelihood. Moreover, settling a hundred "halutsic" families on a hundred dunams did not disallow "Boazite" settlement. "Isn't the country large enough?" Smilansky asked. "Isn't there room for both?" he added in puzzlement. The impediment to large-scale Zionist settlement, he charged, was the ineptitude and inability of the Zionist Executive. Smilansky wound up his case by going on the offensive: "Let's visit reality and ask its opinion: who settled more people per dunam, the 'Boazes' or the 'national' settlement enterprise?" He then answered his own question:

> *Thirty-three* Jewish souls inhabit each dunam in Boazite Petah Tikva and *five souls* inhabit [each dunam] in the "national" settlements of the Jewish Agency. Tel Aviv, the bastion of the Boazes, contains about a third of the country's entire Jewish population. I repeat: if national settlement has made it possible to settle an impoverished family on ten dunams, it is only due to Boazism in the Sharon and the old moshavot.

On this occasion, as in the Ottoman era, Smilansky expressed his faith in the strength of private enterprise and capital as the main engines of economic growth in the Yishuv. Smilansky did not deny the ZO some role in this process but, in contrast to the perspective that he expressed at the dawn of the Yishuv, he denied the ZO one specific role: the builder of settlement infrastructure. Apparently in view of the escalating discord between citrus growers and private entrepreneurs, on the one hand, and the Zionist Movement institutions on the other, and due to the entrepreneurs' ability to pursue most of their economic activities with no dependency on Zionist Movement resources (what is more, as the gap widened and the Farmers' Federation gathered strength, this independence expanded into the fields of welfare and education), the economic functions that the growers earmarked for the national sector contracted to two: research and development, as performed by the experimental station, and lobbying with the Mandatory Government (and even there, not in most cases). Increasingly, the growers identified the government as the public sector in their thinking and their economic endeavors.

About a year after his speech about the "Boazes," Weizmann used the same rostrum to attack the citrus growers again for "exploiting" the Arab worker and, in fact, for employing him. This time, Smilansky seized the opportunity to explain his views on the role of Arab labor in his settlement philosophy. These views were a hybrid of two motives, one obviously related to economic and class matters and one markedly "Ahad Ha'amic." From the economic standpoint, Smilansky emphasized the need to prevent the monopolization of the citriculture labor market by Jewish workers. Weizmann should be aware, Smilansky warned, that "there is one force in the world, a powerful and cruel force, and its name is *competition*; whoever fails to withstand it will stumble and collapse, never to rise." Accordingly, the citrus industry should employ Arab workers as well as the more highly paid Jewish ones. Furthermore, if the Yishuv were to avoid "the fractiousness of political parties and factional jealousies and animosities," it would be economically and numerically stronger. Above all, Smilansky maintained, the national enterprise and Zionism, rather than needing to exclude the Arabs from society and the labor market, should make sure to include them:

> We came to this country to build a *Hebrew Yishuv* and our goal is to make that Yishuv as large, strong, and sturdy as possible. This goal makes it incumbent upon us to distribute our resources among Hebrews: to buy their produce so that their production and, concurrently, their numbers will multiply; and to tap their labor so that they may use it to live and multiply. This is *our main goal*. But we are not alone in this land. There are people here who preceded us. Fate has destined us to be their neighbors forever. We cannot alienate ourselves from them by building a Chinese Wall. We are committed to dwelling with them, producing with them, and trading with them. Thus, we cannot boycott them, shun their goods, refuse to hire them, and patronize only produce and labor that is "one hundred percent" ours. Our labor, like our purchases, should be *mixed* but with a decisive Jewish majority. Otherwise, our aim of creating a large *Hebrew Yishuv* in this country will be hopeless.[39]

Most citrus growers at the time had developed a strong and cohesive worldview, as did the leadership and petty officials of the Zionist Movement and, especially, the Zionist Labor Movement. This *Weltanschauung* consisted of two main planks: national consciousness and a capitalistic socioeconomic outlook. Competition and the profit motive, they believed, would lead to economic efficiency and growth. Opponents of this view agreed that the capitalist system would be profitable for capitalists. However, they developed an "anti-citriculturist" viewpoint based both on national reasoning and on rationales culled from the socialist and Marxist schools. Admittedly, citriculture was not just one of many capitalistic endeavors in the Yishuv economy. Land and labor, the essential inputs of citriculture, were principal components in the value system of the nascent Zionist Movement. Return to and cultivation of the soil were considered *sine qua non* for the restoration of "normal" national life among the Jews. Acquisition of land was an additional and no less important condition for the creation of the resurgent nation's territorial base.

Jewish citriculture in Palestine, based on private land ownership, wage labor, and the profit motive, endangered these basic values. The concerns were three: that growers would prefer not to employ Jews in their orchards because non-Jews were willing to work for less; that they would "exploit" their workers, Jewish and non-Jewish alike; and that, since they held personal title to their land, they might sell it to non-Jews. To forestall such dangers, that is, to assure exclusivity in the labor and land markets, three tools were to be used: national ownership of land, Jewish labor, and "self-labor."

The growers' response, as reflected mainly in Moshe Smilansky's writings, was that the term *national* should not be applied to land and labor only but should be viewed from a broader perspective. From their standpoint, a Jewish majority in Palestine could be formed and most of the country's land area acquired only from a position of economic strength, which would be attained chiefly by capitalistic effort. This argument strongly resembles the explanations of scholars who have attempted to fathom the secret of the West's economic success. David Landes, Nathan Rosenberg, and others have traced the basic secret to the separation of economic decisions, which are the prerogative of firms, markets, and market players, from the domain of political decisions, which belongs to politicians and the political system. Political intervention in the economy, Rosenberg and Birdzell assert, leads to inefficient resource allocation and the impairment of economic growth. To obviate such intervention, the West has developed appropriate ideo-social tools. The principle of natural rights—especially the rights to property and freedom—assures maximum efficiency in economic activity. These scholars, like Smilansky, claim that individuals who pursue their personal welfare amidst personal freedom will do their best to behave with economic efficiency as well. What is more, they will apply their economic initiative time and again. The "cruel judge," the market—Rosenberg, Landes, and others maintain—punishes the negligent and bestows handsome rewards upon those who are efficient and, especially, those who display initiative and verve. What is more, according to this worldview, the attainment of individual welfare enhances the welfare of society at large. The West's path to prosperity, they argue, proves it.[40]

Indeed, the growers saw no contradiction between their personal welfare and that of society and the nation. As the period at issue in this book drew to a close, the Jewish economy of Palestine, including the Labor sector's agricultural activities, underwent processes that made it a close approximation of a capitalistic model. Nevertheless, the derogatory antisocial and antinational image of the private citrus plantation—and of the private economy at large—would persist for many years to come.

PART II

Social and Geographical Platform

THREE

Spatial Distribution and Social and Entrepreneurial Profile

THIS CHAPTER DESCRIBES the spatial distribution of Jewish citriculture from its modest beginnings in late-nineteenth-century Petah Tikva to its expansion across large portions of Palestine in the 1920s and the 1930s. It also attempts, on the basis of the available sources (which are limited, for the most part) to sketch a social and, mainly, entrepreneurial portrait of the Jewish growers. The economic reasons for the expansion of Jewish citriculture and, in the main, the technological and marketing methods that the growers used to sustain the tremendous increase in citrus planted area and yields are discussed in detail in the chapters that follow.

SPATIAL DISTRIBUTION OF JEWISH CITRICULTURE UP TO 1914

Citrus fruit has been grown in Palestine since antiquity. A comprehensive and interesting study by Shmuel Tolkowsky (a leading figure in the industry; his story is told below) about the history of Citrus Fruits in world civilization shows that various types of citrus fruit were grown in the vicinity of Jaffa from the Second Temple era (fourth century BCE–second century CE) to the onset of modern Jewish settlement in Palestine. In 1852, the Dutch researcher and naval officer C. W. M. Van de Velde, who visited Palestine in 1851–1852 and mapped the country, reported that "the oranges of Jaffa . . . appear on the tables of kings in Western Europe."[1]

Thus, citriculture (especially the Shamouti orange) seems to have become a steadily developing export industry in the economy of Palestine by the middle of the nineteenth century. The hub of the industry, as stated, was Jaffa. Demand for the "Jaffa orange" and the introduction of steam-powered vessels in maritime trade in the Eastern Mediterranean—an event that also occurred in the middle of nineteenth century—contributed to the growth of plantations in the environs of Jaffa. The fruit was marketed via Alexandria until 1895, when direct steam shipping was established between Jaffa and the main market for Palestine oranges, Britain. Reflecting the industry's growth, the number of crates of oranges exported from Jaffa increased steadily, from 260,000 in 1895 to 500,000 ten years later.

The success of citriculture did not escape the attention of the country's Jewish inhabitants, several of whom acquired citrus groves in the Jaffa area even before the First Aliya. Just the same, Jewish citrus holdings were exceedingly small in the 1890s.

One of the first Jewish growers was Aharon Leib Fellmann. Fellmann's father, Dov David Halevi Fellmann, acquired in 1883 a forty-dunam orchard in the village of Sumeil, not far from Jaffa. The elder Fellmann passed away shortly after he settled in Palestine, but his widow and sons continued to work the orchard for many years. Other Jewish-owned orchards at the time were situated in Wadi Hanin (subsequently Nes Tsiyyona), Petah Tikva, the Mikve Yisrael school, and in additional localities. Notably, these plantations were either purchased as producing orchards or were planted and developed in accordance with local Arab methods.

Two main factors determined the location of the Palestine orchard in the late nineteenth century: availability of groundwater and proximity to the Mediterranean coast. The first of these factors and, especially, the pumping technology related to it, are discussed separately in the chapters to come. Here we focus on the latter factor, proximity to the seashore or, to be more precise, to the seaport. This was one of the decisive matters in determining the spatial distribution of Palestine citriculture throughout the period discussed in this book.

Since citriculture was mainly an export industry—domestic consumption could hardly sustain it—proximity to the seaport did much to determine an orchard's profitability. What mattered most was the cost of hauling the fruit from the orchard to port on camelback. In his book *Ma'ayan Ganim: Rules for Planting Gardens and Citrus Orchards in the Holy Land* (1891), Aharon Fellmann urged Jewish investors who wished to plant citrus orchards in Palestine not to do so more than "three hours' travel [from the anchorage] or four at the most. . . . One who plants farther away," Fellmann warned, "will find his gains wiped out by his losses, since . . . the farther away one is, the more expensive haulage on camelback to the coast will be and the more purposeless expense one will incur."[2]

Thus, the economic conditions of late-nineteenth-century Palestine conformed with a rule enunciated in 1826 by the agricultural economist Johann Heinrich von Thünen in his book *The Isolated State*. The structure of a farm, von Thünen said, depends mainly on its distance from its market. In von Thünen's basic model, one city serves as an exclusive market for farm produce, means of transport are limited, and the soil structure, level of technical knowledge, population density, and standard of living are homogeneous in all parts of the city's agricultural hinterland. Under these conditions, von Thünen showed that the farther a farm is from its market (in concentric circles that radiate from the city), the more extensive the crops raised and the cultivation methods used to raise them become. The main reason, he concluded, was the increase in haulage cost. Although von Thünen and other agro-economy scholars who followed his lead refined the basic model, von Thünen's principled insight, which treated distance from the market as a matter of crucial economic importance, is accepted to this very day.[3] Accordingly, proximity to Jaffa port (a "market," according to the von Thünen model) was one of the most crucial factors in the spatial distribution of Palestine citriculture in the late nineteenth century.

This distribution did not change until the First Aliya moshavot were established and brought under the influence or patronage of Baron de Rothschild's administration. Citriculture was not central in Rothschild's agricultural program; the leading industry in the moshavot—even in those that were not under his patronage but only under his influence, such as "independent" Rehovot—was viticulture for wine. However, as Moshe Smilansky noted correctly, Baron de Rothschild played a quasi-governmental role in the Jewish settlement enterprise at the time. He invested not only in human capital (the administrative system and the agricultural experts who were associated with it) but also in infrastructure (mainly internal roads but also water systems in the moshavot), capital goods (farm implements such ploughs, pumps, motors, etc.), and education. What is more, in moshavot to which he extended full or partial patronage, Rothschild established experimental farms where his agronomists performed various pilot projects. Some of these experts, such as Peretz Pascal, specialized in citriculture and worked in this field in the service of the baron and the ICA. A few of them, such as Pascal, Meir Apfelbaum, and Abraham Brill, even became private farmers, large-scale private entrepreneurs, and very prominent figures in Jewish citriculture in the late Ottoman and early Mandate eras.

However, the citrus plantings of the baron's experts were very small in land area and were intended for experimental purposes only. Not until the early 1890s did Rothschild's administration begin to plant larger orchards as part of its effort to diversify agriculture in the moshavot. Even then, the extent of this activity was quite limited. By 1900, Petah Tikva had 260 dunams of orchard under the control of the baron's administration. Even after 1900, when Baron de Rothschild revised his policy of subsidizing the moshavot and turned over their management to the formal authority of the ICA (although he continued to pull the strings), his administration continued to provide agronomic know-how, initial investments in new technologies, and financial assistance. The last-mentioned sometimes took the form of subsidized loans to growers, such as a loan to Pardess, the first marketing cooperative of Jewish citrus growers. The loan to Pardess, given in 1903, the founding year of the cooperative, was quite generous for the time: 60,000 francs at only 3 percent (as against the accepted interest rate of 8 percent) for thirteen years, backed by a mutual guarantee from the members of the cooperative.[4]

Unlike viticulture in the moshavot, citriculture was from the very start a markedly capitalistic industry, developed by private entrepreneurs. In Petah Tikva, the cradle of Jewish citriculture as a private-initiative activity, the first two privately developed orchards were planted in 1894, only two years after the baron's administrators established the first citrus orchard in the locality. By 1900, as stated, the baron's apparatus administered 260 dunams of orchard, but these were only 30 percent of the total orchard area (849 dunams) that had been planted in the "mother of moshavot" by that time. Thus, most of the citrus area, 70 percent, was privately owned. Eleven years later, the ratio tilted even more sharply in favor of the private growers. According to 1911 data cited by Curt Nawratzki, who visited Palestine that year to conduct his comprehensive and statistically copious study of the Jewish community in Palestine, the total orchard area in Petah Tikva had grown to 4,806 dunams. Thus, although the area under the control of the Rothschild administration was almost unchanged in absolute terms (at 300 dunams), its share in total orchard area in Petah Tikva fell to 5.4 percent.[5]

The centrality of Petah Tikva in private Jewish citriculture at the time was reflected not only in the large proportional supremacy of private orchards relative to those owned by the ICA in the moshava itself but also in this locality's dominance relative to citrus area in the other moshavot. Thus, the 4,806 dunams of citrus orchards that Nawratzki found in Petah Tikva in 1911 were 68 percent of the 7,038 dunams of Jewish-owned orchards in the country at large.

Table 3.1 underscores a point made above: The rapid growth of private Jewish citriculture, especially that in Judea and Petah Tikva, began after 1900, when Baron de Rothschild curtailed his direct patronage of the moshavot. The main reasons for this were economic. First, as Nachum Gross has noted, the economy of Palestine embarked on a process of growth in 1900 that was influenced only partly by the Second Aliya and its concomitant inflow of human and other capital. Second, the citrus industry itself made an important contribution to this growth. Demand for Palestine oranges increased and technical improvements in local citriculture allowed activity to expand to areas far from Jaffa.[6] Furthermore, one cannot dismiss the claims of Ahad Ha'am, Smilansky, and others, as described at length in the previous chapter, that the abolishment of Rothschild's direct patronage helped to stimulate free enterprise and the growth that it engendered. However, we still

TABLE 3.1
Jewish-Owned Citrus Areas, Early Twentieth Century
(in Turkish dunams =0.919 metric dunam)

	1900		*1911*		*1914*	
Location	*Dunams*	*Percent*	*Dunams*	*Percent*	*Dunams*	*Percent*
Petah Tikva	849	40	4,806	68.3	5,920	62.4
Rishon Lezion	83	4	254	3.6	320	3.4
Nes Tsiyyona	150	7	945	13.4	unknown	—
Mikve Yisrael	161	8	unknown		unknown	—
Rehovot	—	—	504	7.2	2,090[a]	22.0
Total, Judea	1,243	59	6,509	92.5	8,330	87.8
Hadera	179	8	277	4.0	970	10.2
Zikhron Ya'akov	234	11	234	3.3	—	—
Yesud Hamaala	269	13	unknown	—	—	—
Jordan/Jezreel Valleys	—	—	—	—	170	1.8
Others	199	9	18	0.2	20	0.2
Total	2,124	100	7,083	100.0	9,490	100.0

Sources: 1900—Dov Gurevich and Aharon Gerz, *Jewish Agricultural Settlement in Palestine* (Jerusalem: The Jewish Agency, Department of Statistics, 1938) (Hebrew), 43*, Table 31; 1911—Curt Nawratzki, *Die Juedicshe Kolonisation Palaestinas* (Muenchen: Reinhardt, 1911), 338; 1914—Gurevich and Gerz, *Jewish Agricultural Settlement*, 44*, Table 32; David Yudilowitz (ed.), *Rishon Lezion, 1882–1941* (Rishon Lezion: Carmel Mizrahi, 1941), 255 (Hebrew).

a. Includes Nes Tsyonna and Mikve Yisrael.

must explain the centrality of Petah Tikva and Judea, as opposed to other places, in the spatial distribution of Jewish citriculture at this time.

The explanation seems to lie in several factors. First, the light soil structure and temperate climate of Judea made the moshavot in this area better suited to citriculture than moshavot farther north, such as Zikhron Ya'akov and localities in the Galilee. Second, the distance from the main port of export, Jaffa, was a decisive factor in the development of citriculture in the Jewish moshavot, especially in Judea. In other words, von Thünen's aforementioned model of concentric circles may explain the concentration of citrus in Petah Tikva and, later on, in moshavot such as Nes Tsiyyona and Rehovot. However, additional social and cultural factors related to the separate development of each moshava should also be taken into account.

A good example is the absence of citriculture development in Rishon le-Tsiyyon at that time, even though this moshava was quite close to Jaffa.

There seem to have been two main reasons for this: a low groundwater table and the development of viticulture. Rishon Lezion is famous in that no water (even potable water) was found there until Baron de Rothschild brought in experts and drillers who eventually located some at the daunting depth of forty-two meters. Apparently the groundwater table in the Rishon Lezion area was uneven. Wells were sunk to depths of between twenty and twenty-five meters and several orchards were planted around the moshava (especially near springs). However, significant development of citriculture in Rishon Lezion did not occur until the late 1920s. As I show below, the expense of installing an irrigation system was an important factor in the total investment in the preparation of an orchard. Clearly, the greater the investment needed to acquire the knowledge to find water (since the depth of the groundwater table was unknown) and to sink a deep well, the costlier the total investment would be. Accordingly, these factors impeded the development of citriculture in this particular locality at that time.

The other reason, as stated, was the centrality of viticulture in Rishon Lezion and the large winery that had been built there. Thus, by the beginning of the twentieth century, when demand for citrus surged, Rishon Lezion had already become the center of viticulture in the moshavot south of Tel Aviv. Additionally, Rishon Lezion was a regional administrative center during the era of Rothschild's patronage. These factors, by reinforcing the existing economic and socioinstitutional conditions, sustained the primacy of wine viticulture in the agriculture of Rishon Lezion. In 1911, Rishon Lezion had about 4,000 dunams of vineyards as against 254 dunams of citrus. In Petah Tikva, in contrast, there were 4,806 dunams of orchards and 1,000 of vineyards. By 1914, things had not changed much: Petah Tikva had 5,920 dunams of citrus and only 1,340 dunams of grapes, whereas Rishon Lezion maintained its centrality in viticulture and did not develop its citrus. In 1918, after the war and the planting moratorium that its dire economic conditions caused, Rishon Lezion had only 320 dunams of citrus as against 5,054 dunams of vineyards.[7]

At the beginning of the twentieth century, Petah Tikva, unlike Rishon Lezion, had abundant water only 3–10 meters from the surface and its proximity to Jaffa port—about seventeen kilometers, according to Nawratzki—stimulated the development of citriculture. However, since there were no paved roads from the moshavot and orchards to Jaffa and hardly any motor vehicles in the whole country, haulage was done on camelback. There were horse-drawn carts, but they tended to break down frequently due to the state of the roads. A camel could haul roughly the same load as a cart hitched to one horse (three hundred kilograms or so) but its maneuverability in dunes and mud gave it an edge over the equine alternative. However, haulage by camelback was expensive. The expense of hauling a crate's worth of oranges from Petah Tikva to Jaffa was 8 percent of the total expense in preparing the crate for export: materials, picking and packing, haulage, taxes, and loading

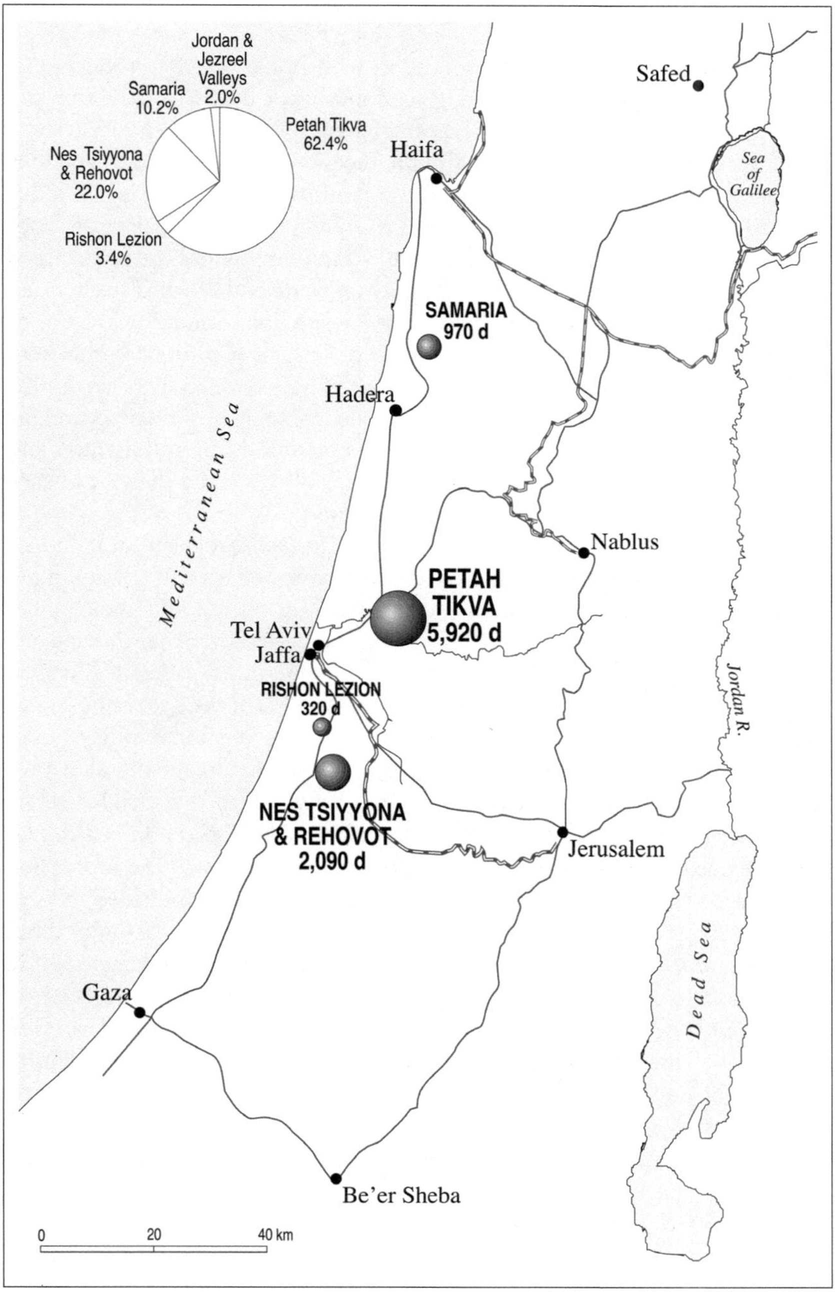

Map 3.1. Spatial distribution of Jewish citriculture, 1914.

onto the vessel. If the moshava was far from Jaffa, as in the case of Rehovot, haulage was even costlier. Furthermore, since Petah Tikva was the farthest of the Judean moshavot from the winery in Rishon Lezion, the produce of its vineyards was priced higher than that of moshavot closer by. It is true that the camel caravans hauled the grapes of Petah Tikva to Rishon Lezion at night to keep the fruit fresh, but the produce was degraded by the bouncing that it suffered along the way. Thus, the farmers of Petah Tikva had a dual incentive to plant citrus and abandon the viticulture that Baron de Rothschild had established: the distance from Rishon Lezion and the proximity to Jaffa. This was especially so after 1900, when Rothschild turned his ventures over to the ICA and the subsidization policy was discontinued.

It is proper to stress the importance of Jaffa as a center for the citrus industry as the country's principal export port and commercial town, a place where growers could call on agents of knowledge, sources of finance, and factories that sold mechanical implements and offered repair and maintenance services. The proximity of Petah Tikva to Jaffa also lowered the cost of these services that the city provided to the hinterland.

The ascent of Petah Tikva to centrality in Jewish citriculture, and the economic and social growth that citriculture bestowed on this colony in particular and the Zionist colonies of Judea generally, were points that Smilansky stressed in his debate with Ruppin and the Labor Movement ideologues, as noted. Indeed, the steady increase in the populations of Petah Tikva and the Judean moshavot in 1900–1914—a consequence of the economic success of these localities, in which the role of citriculture was increasingly important—stood out in the demographic and economic landscape of Palestine at the time. Indeed, the population of Petah Tikva more than tripled in the fourteen years following the elimination of Baron de Rothschild's patronage. On the eve of World War I, the "mother of moshavot" was the most populous Jewish agricultural locality and contained about one-fourth of the Jewish agrarian population at large. Rishon Lezion maintained its importance, mainly due to the centrality of viticulture in its farming endeavors. In an interesting contrast, the population of Zikhron Ya'akov, like Rishon Lezion an administrative center of the baron's administration, failed to grow. Thus, its share of the Jewish agrarian population in 1914 declined perceptibly. Finally, the population of two famous settlements of the Labor Movement—the first kibbutz, Degania, and the training farm, Kineret, accounted for only 0.7 percent of the agricultural population of the Yishuv at the time.[8]

THE FIRST JEWISH GROWERS— ENTREPRENEURIAL AND SOCIAL PROFILE

What kind of people were the first Jewish citrus growers in Palestine? Below, we attempt briefly to retell the stories of several of them and sketch their socioeconomic and entrepreneurial profile in view of the major attributes listed in chapter 1.

The first on our list is *Shimon Rokach* (1863–1922), the offspring of a Jerusalem family and a descendant of the famous Jewish printer in Palestine, Israel Bak (1797–1874). In 1884, the Ottoman authorities awarded Rokach's father (and several Arab dignitaries in Jerusalem) a toll franchise on the Jaffa-Jerusalem road, and young Shimon was sent to run the business from Jaffa. There, he helped to found a new Jewish neighborhood (Neveh Zedek), chaired the united committee of Jaffa residents, and founded various welfare and mutual-assistance societies. In 1886, among his other activities, Shimon Rokach, together with his brother Elazar Rokach (who at the time was secretary of the Hovevei Tsiyyon committee in Jaffa) and members of the First Aliya established Ezrat Yisrael, a society that aided newly landed Jewish immigrants in Jaffa and helped farmers and laborers in Petah Tikva. When the planting of citrus orchards in Petah Tikva began, Rokach established two large orchards in conjunction with two affluent friends in Jerusalem, Yehezkel Blum and Leib Levy. In 1905, he joined up with Blum, Levy, his brother-in-law Shmuel Moyal (son of the Hovevei Tsiyyon agent in Jaffa, Avraham Moyal), and two of Baron de Rothschild's agricultural experts who had since gone into private farming, Meir Apfelbaum and Peretz Pascal, to acquire land for, and to plant, the largest Jewish-owned orchard up to World War I, the 650–dunam Bacharia orchard on the banks of the Yarqon River.

However, Shimon Rokach's most notable entrepreneurial accomplishment was his role as the living spirit behind the founding and, for years, management of the Pardess cooperative. Until his death in 1922, he provided Pardess with strong leadership and moved it in innovative directions amidst continual risk. It was not only in the very act of founding the Yishuv's first cooperative society of its type that Rokach stood out as an innovative entrepreneur who operated under conditions of uncertainty. This quality was also evident in other undertakings of Pardess under Rokach's leadership. Rokach acted to establish a eucalyptus sawmill to make boards for orange crates (World War I nipped this venture in the bud). Soon after Pardess was founded, Rokach moved its offices to the port of Jaffa so that he could more effectively oversee its commercial activities. At about the same time, Rokach, at considerable risk to the shipments of the members of his cooperative, concluded an initial contract with a British shipping firm to circumvent the Arab traders who had controlled this market until then. Contemporary sources also note that at times of falling profits, Rokach applied effective leadership vis-à-vis his associates and encouraged them to continue sustaining the cooperative that they had set up. He also regularly sought new markets for Pardess fruit, from Australia to the Far East.

Rokach's actions indicate that his worldview was clearly nationalistic, albeit not in the manner the Zionist Labor Movement. The thread that connects his activities conspicuously is ceaseless effort to strengthen the society and economy of the Yishuv, in which he was a member of the elite. Study of the minutes of Pardess board meetings shows that not only Rokach but also many members of the cooperative held solidly nationalistic views, which they took for granted, and worked toward the practical goal of strengthening the Yishuv economy.

One may characterize Shimon Rokach as the embodiment of many of the entrepreneurial attributes listed in chapter 1. First, he was a "Schumpeterian" entrepreneur who blazed new trails in organizing, marketing, and changing the economic and social equilibrium of his society. He was also Schumpeterian in that even his contemporaries considered him a leading and almost heroic figure. Second, Rokach qualifies as an entrepreneur if only because he acted continually under conditions of uncertainty, took considerable risks, made strategic economic decisions, and sensed new opportunities that others overlooked. Obviously, he was a *private* entrepreneur since he used private capital (including his own), in his activities, and since the first and main purpose of his initiatives was private, that is, to meet the needs of other entrepreneurs and not those of an anonymous collective.[9]

Another prominent personality among the first growers in Petah Tikva was *Dr. Aharon Meir Mazie* (1858–1930), whom Baron de Rothschild had appointed doctor of the southern moshavot. Mazie, a litterateur, a linguist, an alumnus of Mir Yeshiva, and a rabbi who had acquired his ordination at Esriel Hildesheimer's neo-Orthodox Rabbiner Seminar für das Orthodoxe Judentum in Berlin, took medical training at Rothschild's urging and specialized in Paris in ophthalmology. In 1888, after the "well-known philanthropist" (as Rothschild was known) named him chief physician of the Judean moshavot, he moved to Palestine. Mazie spent twelve years based in Rishon Lezion, the administrative hub of Baron de Rothschild's colonies south of Jaffa. There he directed the first hospital in the moshavot and served as the regional medical inspector. When the ICA took over in 1900, Mazie and his family moved to Jerusalem. The sources do not identify the provenance of Mazie's wealth but leave no doubt that it was vast. In Jerusalem, he lived in a mansion on Mesilat Yesharim Street, not far from the German mission Talitha Kumi. He shared the residence with his wife and with the family of his son-in-law, Zvi Izakson (another leading citrus grower in Mandate-era Palestine). In the late 1920s, a small modern school was established in the building, where some thirty pupils took a curriculum based on the Religious Zionist doctrine. In Jerusalem, Mazie served as chief physician of Bikur Cholim Hospital.

As noted, Mazie was not only a doctor but also an author and a linguist. Immediately after he settled in Jerusalem, he collaborated with David Yellin and Eliezer Ben-Yehuda in the establishment of the Hebrew Language Committee. In 1922, when Ben-Yehuda died, he took over the presidency of the committee and held this post until his own death eight years later. Mazie coined many new Hebrew words, foremost in medicine, and corresponded extensively with people who sought his authoritative views on the subject. The culmination of his comprehensive work on the topic—*Lexicon of Hebrew Medicine and Natural Science Terms* (1934)—was published posthumously, and the editor who applied the finishing touches to Mazie's research was the Hebrew poet (and a physician on his own right) Shaul Tchernichovsky.

Mazie took part in additional activities that enhanced Jewish national consciousness: He was extensively involved in the endeavors of the Palestine Exploration Society (subsequently the Israel Exploration Society) and was its treasurer. He was also a signatory to the founding scroll of the Hebrew University of Jerusalem, an institution that made an important contribution to the strengthening of modern Jewish national consciousness.

Since only Ottoman subjects were allowed to hold title to land at the time, Mazie, as an Ottoman subject, was one of two trustees (David Yellin was the other) of the Ahuzat Bayit society, which established the neighborhood that would soon give rise to Tel Aviv. When construction of the Ahuzat Bayit neighborhood began in 1909, Mazie was one of the first purchasers of a plot there. In 1925, he was a partner in the construction the largest and most important hotel in Tel Aviv to that time, the Palatin, on the plot that Mazie had acquired from the Ahuzat Bayit sponsors.

Mazie's roles as a doctor, a linguist, an activist in research of Jewish and Palestinian culture and history, and a trustee for the founders of Ahuzat Bayit are somewhat known. His entrepreneurial activities in citriculture, in contrast, are shrouded in obscurity. In fact, Mazie was one of the first citrus growers in Petah Tikva and a prominent personality in the industry at large. In 1896, while in Rishon Lezion, Mazie acquired one hundred dunams in Petah Tikva and planted an orchard there. Several years later, he participated in the founding of the Pardess cooperative. Shimon Rokach was the society's powerful and influential director; Mazie was its chairman until his death in 1930.

Thus, several characteristics of the entrepreneur were integrated in the persona of Aharon Meir Mazie, as they were in Rokach. Mazie broke new ground in organization and marketing or, at the very least, was an active partner in the strategic decisions that led to these initiatives. He acted prodigiously under conditions of uncertainty and, like Rokach, held strong nationalist views. Several of his ventures (like Rokach's) were in the public domain—the Language Committee, the Hebrew University, the Exploration Society—but concurrently he took private initiatives, for example, in Ahuzat Bayit, in hostelry, and in citriculture. Notably, Mazie, like Rokach, did not live on his farmstead; he belonged to the class of "absentee landlords." Finally, for Mazie, like Rokach, citriculture was an important but not a sole business activity.[10]

One may use the incomplete available sources to sketch a common social and economic profile of the first citriculturists in Petah Tikva. Such an inquiry brings several interesting conclusions to light. (1) More than 40 percent of the growers were "absentee landlords." Some lived abroad; others dwelled in Jaffa, Jerusalem, or (in the case of Dr. Mazie), Rishon Lezion. This finding reinforces the capitalistic image of Jewish citriculture at its outset. (2) Some 54 percent of investors in citrus orchards were born outside of Palestine. This comes as no surprise, since Jewish settlement in Palestine at the time was an immigrant enterprise. The surprising fact is that 27 percent of the investors, such as Rokach, Moyal, and Apfelbaum, were natives of Palestine. (3) Quite a few (31 percent) were employed in Baron de Rothschild's administration before or during their activity as growers. Subtracting the absentee landlords (including Dr. Mazie, who worked for the administration but did not cultivate his own orchard), we find that seven of the twelve growers who lived in Petah Tikva (58 percent) worked or had worked for Rothschild. This provides further support for a contention previously broached—in the research literature and in this study—about the crucial nature of Baron de Rothschild's activity in creating an infrastructure for private initiative.

By working for Rothschild, Apfelbaum and Pascal gained agronomic knowledge and earned income that they used in their business ventures. When he went into private farming, Pascal even received as severance pay part of the orchard that he had managed for Rothschild. Others, such as the surveyor Moshe Slor, treated their work in the Baron's administration as a secure source of livelihood until their investments could mature. However, we emphasize again that overall, including the absentees, two-thirds of the early citriculturists in Petah Tikva were able to support themselves and invest in their orchards without Rothschild's assistance.[11]

As stated, although Petah Tikva was the center of citriculture at the time, the industry began to develop elsewhere as well and entrepreneurs who powered its expansion could be found in these localities, too. Noteworthy among them is an almost anonymous figure, *Ephraim Sachs* of Rehovot. Sachs attempted to settle in Palestine in the 1890s but the Turkish authorities thwarted the move. He emigrated to the United States and settled in Chicago. In 1907, after establishing himself and amassing wealth, Sachs returned to Palestine (over the objections of his wife and daughter) and established his home in Rehovot. Together with his son-in-law, Tovia Miller, he became a leading personality in the moshava and one of its first citriculturists. He managed the local branch of the loan cooperative that the Anglo-Palestine Bank had established in 1912, organized a central postal service for the moshava, and lent the moshava 25,000 francs to install a running-water system. Sachs was also chairman of the municipal council of Rehovot for much of the Mandate era.

Another pioneering citriculturist in Rehovot was *Tovia Miller*, Sachs's son-in-law. Like Sachs, Miller was peripatetic, migrating from his birthplace

FIGURE 3.1. Tovia Miller and Moshe Smilansky in front of Miller's house in Rehovot.

in Eastern Europe to South Africa and then to New York before settling in Rehovot in 1909 and planting an orchard with his father-in-law. Miller was one of those growers who took a regular interest in all technological innovations in the industry and often tried to adjust them to local conditions and apply them in his orchard. During the Mandate era, Miller co-edited the farmers' journal *Bustenai* (along with Smilansky) and was in charge of the journal's professional affairs. Sachs and Miller could not have been prompted to move to Palestine for reasons of material distress; their relocation obviously points to national consciousness.[12]

Another important personality in early Jewish citriculture who displayed a clearly nationalistic mindset was *Shmuel Tolkowsky*. Tolkowsky, born to a family of diamantaires from Antwerp, preferred not to stay in the family business and decided instead—acting on the Zionist worldview that he had developed in his youth—to study agronomy in order to apply this knowledge to the conditions of Palestine. Tolkowsky participated in the IX Zionist Congress (1909) and settled in Palestine in 1911. The Palestine Office officials suggested that he purchase a parcel in the Jezreel Valley but Tolkowsky, seeing no economic future in private settlement in the valley (most of which had not yet been acquired), moved to Rehovot and bought the orchard of Zalman Minkow, who died in 1912. (Minkow had been the first citriculturist in this moshava, having planted his orchard in 1904.) Shortly after he made the purchase, Tolkowsky was among the powers behind the founding of a rival citrus marketing organization to Pardess, the Merkaz cooperative. During World War I, Tolkowsky found himself in Britain, where he was active in the Zionist circle that negotiated the wording of the Balfour Declaration. Tolkowsky married the daughter of the Zionist mogul Isaac Leib Goldberg (the "unknown benefactor"). Returning to Palestine, Tolkowsky settled in Tel Aviv, joined the Hagana central committee, and continued to play important roles in citriculture. He managed the large orchard that his father-in-law had bought (the "Goldberg orchard" in Ramat Gan); was a partner in a citrus marketing venture called the Jaffa Fruit Company; managed for some time the orchards of Gan Haim, a firm in which the partners were Alfred Mond, Chaim Weizmann (for whom the company was named), and others; and, in the main, directed the Jaffa Citrus Exchange, the umbrella organization of most Jewish growers during the Mandate era (about which we discuss at length below).[13]

Shmuel Zvi Holzmann (1883–1960) was a citriculture entrepreneur of a totally different type than Tolkowsky, although both began their activity in the industry in Rehovot. Holzmann was born in Jerusalem to a family of vintners. At the age of sixteen he left Palestine to attend a rabbinical academy in the United States. However, he quickly gave up this vocation and headed for South Africa, where he raised ostriches. In 1906, after his ostrich business earned him a small fortune, Holzmann returned to Rehovot, bought a fifty-dunam parcel, and planted an orchard. Holzmann was one of the first citriculturists in this moshava. His uniqueness as an entrepreneur lay in the variety of his initiatives. They include an attempt (several years after he returned from South Africa) to breed ostriches in Palestine, large-scale planning of settlement activity in the 'Arava (the desert area between the Dead Sea and the Gulf of 'Aqaba), mineral exploration in the Negev, and the establishment of a mountain agriculture village and health center in the settlement known to this day as Kefar Etsiyyon ("Etsiyyon" being a Hebrew rendition of Holzmann's name). All these ventures ended in failure, and the Kefar Etsiyyon enterprise saddled him and his investors with a considerable financial loss.

As he went about these ventures, Holzmann imported botanical knowledge and introduced modern technology in citriculture and for other uses. In citriculture, Holzmann attempted to introduce new varieties (such as the pomela and the tangerine) and to make orchard irrigation more efficient (by importing sprinkler systems), and he made the knowledge in his possession available to other growers, who corresponded with him extensively. The Jewish Agency's Agricultural Experiment Station in Rehovot maintained close relations with Holzmann, who allowed it to use his orchard for citrus experiments. Holzmann's private archives contain a list of fifty-nine diverse innovations that he introduced in Palestine, nearly all of which were related to agriculture.[14]

Thus, Holzmann was a "Schumpeterian" entrepreneur of the type that Fritz Redlich noted—one who does not succeed in all his/her ventures but whose innovative enterprises make a long-term contribution. As one of the vanguard citriculturists in Rehovot, he definitely took risks and broke new ground in the ventures that he pursued at the time.

An important family of growers in early Jewish citriculture were the *Jacobsohns*, originally from Zhagare, Lithuania. In the 1860s, after the reformist Czar Alexander II allowed Jewish merchants of the "first guild" (the highest class of taxpayers) to settle outside the Pale of Settlement, the merchant Shraga Feibl Jacobsohn and his family moved to Moscow, where he and his son Mordechai did a brisk trade in knitwear. The Jacobsohns were one of the wealthiest families in Russian Jewry and belonged to a circle of affluent Jews such as Kalonymus Zeev Wissotzky, who also came from Zhagare and reached Moscow shortly before the Jacobsohns did. Wissotzky was also one of the first to join Menuha ve-Nahala, a confraternity established on February 27, 1890, in Warsaw at the initiative of the members of Benei Moshe (a secret order of Hovevei Tsiyyon) in the Polish capital. The goal of Menuha ve-Nahala was to establish a new moshava in Palestine with its own resources and without any subsidies whatsoever. A year after the founding of Menuha ve-Nahala, Mordechai Jacobsohn and his family moved to Palestine and settled in the moshava that the society had established, Rehovot. The family patriarch, Shraga Feibl, followed them to Palestine two years later and settled in Jerusalem.

In 1907, pursuant to Rothschild's policy (implemented by the ICA) of cutting back on vineyard area in the moshavot, about forty grape growers in Rehovot were given "uprooting compensation" in return for an undertaking to use the money to plant orchards. Only three of the forty—Eliezer Jacobsohn (Mordechai Jacobsohn's son), Moshe Smilansky, and Y. Feinstein—decided to plant citrus; the rest planted almonds. This marked the beginning of citriculture in Rehovot. By World War I, *Eliezer Jacobsohn* was one of the most prominent personalities in private citriculture. His prominence increased after the war, when he served with Ephraim Sachs as a member of the Loan Fund board and was a leading figure in the Farmers' Federation and the Pardess cooperative. As a resident of Rehovot, he also managed the orchards of his relatives and of other "absentee landlords" who lived elsewhere in Palestine or abroad.

Eliezer's brother, *Zalman Jacobsohn*, was born in 1879 in Moscow. After moving to Palestine in 1891, he attended his grandfather's yeshiva in Jerusalem and later went to Paris to study Russian literature at the Sorbonne. He abandoned this calling shortly after the beginning of his studies and immersed himself in commercial and entrepreneurial activity. In 1904, he returned to Palestine and went to work for the Carmel company as a marketer of the wines of Palestine. A short time later he went abroad again, this time as a representative of the Joint Association of Rishon Lezion and Zikhron

Ya'akov Vintners. From then until 1912, Jacobsohn stayed abroad apart from occasional visits to Palestine. In 1907, during one of these visits, he married Rosa, daughter of Zalman David Levontin, a founding member of Rishon Lezion and the director of the Anglo-Palestine Company (subsequently the Anglo-Palestine Bank) in Palestine. Contemporary documents show that while he was abroad Zalman Jacobsohn went into the citrus trading business. Many documents, especially from the 1911/12 season, show that he spent that season in Britain and traded in oranges while serving as the representative of the Anglo-Palestine Bank.

Returning to Palestine in 1912, Zalman Jacobsohn began a new chapter in his economic activity in Palestine citrus. In a manner related to his association with the Anglo-Palestine Bank but on his own, he began to deal in Palestine citrus. He established in Jaffa the first Jewish-owned private office for citrus exports and exported the produce of both Arabs and Jewish growers (including Moshe Smilansky). When World War I began, Jacobsohn's wife and children, along with his in-laws the Levontins, left Palestine and spent the next three and one-half years in Egypt. Zalman himself stayed in Palestine; in 1916, when the Turks ordered the inhabitants of Tel Aviv to evacuate, he moved to Tiberias and joined the "migration committee" that helped those who relocated to the north of Palestine. After the British occupied the country, he served for a year and a half as an inspector with the official Zionist delegation, sent to Palestine to help Britain to organize there an autonomous Zionist community. Concurrently, he set up an import-export enterprise in a partnership with Meir Dizengoff that fell apart about a year later. Precisely then, however, Jacobsohn's businesses, especially those involving citrus, began to expand significantly and made him into a key figure in citriculture. During the Mandate era, Zalman Jacobsohn held important positions on various government committees that dealt with citrus. In 1929, after the Arab uprising of that year, Jacobsohn—in conjunction with Isaac Rokach and Shmuel Tolkowsky—established the Jaffa Citrus Exchange.

Near the end of the period of concern in this study, Zalman Jacobsohn joined the Farmers' Federation and quickly rose to importance. Apart from being a member of its financial and educational committees, he chaired the board of the agricultural school in Pardes Hannah, established at the Federation's initiative. It is also noteworthy, as stated, that as he went about all these activities Zalman Jacobsohn owned orchards in Rehovot.[15]

Finally, it is appropriate in this context to mention again the role of *Moshe Smilansky* in Jewish citriculture at its outset and, especially, his centrality in the industry after the British Mandate era began.

These biographical notes about the activities of some of the most prominent entrepreneurs—and several less prominent ones—in early Jewish citriculture in Palestine may, in my opinion, provide an adequate basis for an answer to a key question: What prompted them to invest in Jewish citriculture? It seems to have been mainly a business decision, since many of them pursued additional private business ventures in Palestine and abroad and quite a few did not live in the moshavot at all. Even those who inhabited the moshavot, such as the first growers in Petah Tikva and Rehovot, could have grown alternative crops—such as vegetables, grapes, or almonds—in their places of residence. Thus, their decision to invest in citriculture was a pronouncedly entrepreneurial one, that is, chiefly a decision made under conditions of uncertainty and, usually, at financial risk. As I have also shown, some of them backed their business decisions with actions that typify the Schumpeterian entrepreneur.

If the entrepreneurial and capitalistic nature of investing in citriculture is undisputed, the extent of these personalities' national motives lends itself to debate. The roots of the debate, as I showed in chapter 2, lie in the suspicion if not hostility of many leading figures in the Zionist Movement (especially, the heads of the Labor Movement) toward private enterprise and capital. I would suggest, however, that the information we have presented thus far about the national motives that underlie the activities of standout personalities such as Smilansky, Tolkowsky, Rokach, and Mazie—and those of secondary figures in the industry such as the Jacobsohns, Tovia Miller, and Ephraim Sachs—suffices to refute the argument, which still passes for conventional wisdom in research, that in Zionism private entrepreneurship and national motives are mutually exclusive.[16] From the standpoint of people who made a willful decision to move to and invest in Palestine when they could have stayed or invested abroad, private initiative was perceived not as contradictory or even neutral to the national outlook but rather as an inseparable, if not central, part of it. Tolkowsky refused to stay in Antwerp and enter the diamond business because he was a Zionist. However, he rejected the possibility of buying land in the Jezreel Valley because he did not considered such land a worthwhile investment. Instead, wishing to make a good living in Palestine, he bought orchard land in Rehovot.

Not only did the entrepreneurial business motive not clash with the national motive; it actually reinforced it. In other words, the main goal of the growers whose biographies are retold here—a group that, in my opinion, provides a rather accurate reflection of those involved in the industry at the time—transcended the maximization of profit and included the wish to obtain a handsome social and financial return on the investment. To their way of thinking, personal well-being complemented and corresponded to the well-being of society at large; the two did not contradict.

SPATIAL DISTRIBUTION DURING THE BRITISH MANDATE ERA

Until early 1927, the spatial distribution of Jewish citriculture was basically unchanged from the pre–World War I pattern, even though the planted area in Jewish hands had more than doubled from 9,500 dunams to 19,600 dunams during the intervening years. In the decade that followed, planted area grew more than eightfold. Thus, the spatial distribution of the orchards also changed, as Table 3.2 shows.

TABLE 3.2
Jewish Owned Citrus Areas, 1927 and 1936

	Dunams		*Percent*	
	1927	*1936*	*1927*	*1936*
Petah Tikva	8,460	14,986	43.1	9.2
Nes Tsiyyona	2,692	9,438	13.7	5.8
Rehovot	2,178	14,906	11.1	9.2
Rishon Lezion	1,181	10,156	6.0	6.2
Others in Judea	2,125	27,486	10.8	17.0
Total, Judea	16,636	76,972	84.7	47.4
Hadera	1,898	8,734	9.6	5.4
Zikhron Ya'akov	200	1,228	1.0	0.7
Total—Samaria	2,098	21,284	10.7	13.1
Total—Sharon Plain	785	59,488	4.0	36.7
Degania and Kineret	120	317	0.6	0.2
Galilee/Jezreel Valley	—	4,189	—	2.6
Total	19,639	162,250	100.0	100.0

Sources: 1927—Melekh Zagorodsky, *The Citrus Book: Manual for Orange Growers, Workers, Experts, and Nurseries Owners* (Tel Aviv: Hahaklai, 1929), Vol. I, 5–9 (Hebrew); 1936—Ben-Israel (Ze'ev Smilansky), "Development of Hebrew Citrus Orchards," *Bustenai* 8: 31–33, 42, 46 (1936/37); ibid., 9, 4, 14, 27 (1937/38) (Hebrew).

Thus, as Table 3.2 shows, even in early 1927 Petah Tikva remained the center of Jewish citriculture. Although its share in total Jewish citriculture declined from 62 percent in 1914 (see Table 3.1) to 43 percent in 1927, no competitors had come into sight. Petah Tikva lost proportional ground mainly due to the growth of other moshavot in Judea, especially Rehovot. The Samaria region, foremost Hadera and the PICA-owned Nazleh orchard near Zikhron Ya'akov, accounted for about 10 percent of the Jewish citrus area. The establishment of a new citrus region, the Sharon Plain, led to a slight change. In moshavot such as Kfar Saba, where the land had belonged to the farmers of Petah Tikva since the Ottoman era, and in new Zionist colonies established at the beginning of the Fourth Aliya, such as Magdiel and Herzliya, planting of citrus began at this time. Although the momentum of this activity would escalate, in early 1927 the process was in its infancy.

Accordingly, Jewish citriculture continued to center in the Judean moshavot. In 1914–1927, the share of this region decreased only slightly, from 88 percent to 85 percent. In contrast, the relative share of citrus area in the Jordan and Jezreel valleys, minuscule to begin with, contracted between 1914 and 1927 from 1.8 percent to 0.6 percent. The Labor Movement settlements, most of which were in the Jezreel and Jordan Valley areas, had not gone into citriculture as of the beginning of 1927.

The slight change in the spatial distribution of citrus areas during the 1914–1927 period, coupled with the doubling of planted area, shows that the most evident planting activity occurred in the vicinity of the moshavot themselves. One may explain this in several ways. First, much of the newly planted area belonged to second-generation farmers who wished to stay in their moshavot. The availability of knowledge, means of transport, labor supply (Arab and Jewish), and sources of finance (mainly savings-and-loan cooperatives that had come into being in the moshavot), along with the social and cultural mass that the veteran moshavot had built up, encouraged farmers to develop new citrus areas near the long-standing moshavot. Indeed, these factors largely dictated that outcome. Importantly, too, the capital accumulated by the veteran growers in the high-profit years enabled them and their families to make additional investments in citrus.

These factors were augmented by the steady increase in the population of the moshavot. Between 1922 and 1927, for example, the population of Petah Tikva nearly doubled, from 3,000 to 5,800. Other moshavot experienced significant demographic increase: Rishon Lezion by 50 percent (from 1,400 in 1922 to 2,100 five years later) and Rehovot by "only" 40 percent (from 1,200 to 1,700). However, since the Judean moshavot were almost eight times as populous as the Sharon area in 1927 (15,000 versus 1,900), even the "moderate" proportional demographic increase in the other moshavot of Judea contributed to the growth of the citrus industry. Indeed, a statistical analysis and detailed description of the social, economic, and demographic makeup of Rishon Lezion, published in 1930 in the journal *Mis-'har Ve-Ta'asiya* (Trade and Industry) reported that "The Fourth Aliya alone boosted [the population of moshava farmers] by 10.5 percent, [and some of the new inhabitants] have acquired very decent-sized parcels and increased the cultivated area greatly." The article also reported on a phenomenon in the moshavot that accelerated until the late 1930s, the replacement of vineyards with citrus orchards. This, too, may explain the increase in citrus area in the old moshavot.[17]

To build up the moshava area, it was necessary to purchase land continually from Arab landowners around the settlements. For example, Moshe Smilansky and Tovia Miller—in coordination and competition with the Yishuv's great land-purchase entrepreneur, Yehoshua Hankin—made regular acquisitions in the vicinity of their moshava. Between 1921 and 1930, Smilansky and other residents of Rehovot (as Smilansky reported in a letter to Hankin) acquired 8,670 dunams from landowners nearby.[18] There was a connection between the moshavot as convenient places to absorb newcomers (in the economic, cultural, and social senses) and their function as appropriate locations for fulfillment of the aspiration to settle the country by working its soil, on the one hand, and the acquisition of land for participants in the Fourth Aliya, on the other. Smilansky expressed the nexus by proposing to Arthur Ruppin in June 1924, at the beginning of the Fourth Aliya, that the Zionist Organization institutions undertake the purchase of land for middle-class immigrants and avail themselves of the services of moshava members in so doing. This, I would say, was one of the first proposals that envisaged comprehensive cooperation between moshava settlers and the Zionist Movement institutions for the settlement of members of the middle class. Thus, Smilansky wrote to Ruppin:

> In the past few weeks, many people from Poland, Lithuania, and even America have been coming to the moshavot and wishing to buy land. They want to buy [it] in the moshavot specifically. The type of people, the living conditions, and the economic conditions in the moshavot suit their tastes. They are not afraid of "exploitation," of course. Some of them are wealthy, others are affluent, and still others have scanty means.
>
> Around the moshavot is land that can be bought from Arabs. Around Rehovot, one can even buy thousands of [Turkish dunams] with regularized legal status at 3–4 Egyptian grush per dunam. . . . I propose that you take over this initial [activity], either by the Palestine Land Development Company or the Zionist Executive. This [institution] should also be interested in immigration of members of the householder class. Insofar as it pertains to Rehovot and its vicinity, I undertake to help you in all your endeavors. And you will find people like me in all the moshavot.[19]

In addition to the conditions noted thus far, one of the main reasons for the spatial distribution of Jewish citriculture during these years, as in the Ottoman era, was the location of the industry relative to its market. In other words, due to improvements in the efficiency and availability of transport, the moshavot of Judea had better connections at this time to the port of export, Jaffa, than other citriculture and agriculture areas.

The transport system of Palestine began to change perceptibly as early as World War I. Most main localities, including the moshavot of Judea, were connected with urban centers by dirt and crushed-stone roads. Furthermore, railroads were built in a grid that linked the southern and northern parts of the country. Notably, however, these changes had little effect on the citrus moshavot in Judea, for several reasons. (1) The railroad bypassed the Jewish settlements. (2) The road system that was built during the war was comprised of dirt roads and roads made of crushed stone. They were difficult to navigate in wheeled vehicles and barely survived the vicissitudes of weather in the winter, the citrus export season. (3) As a rule, Palestine still had few civilian motor vehicles. They did not become common until civilian rule was established in July 1920 and the new regime began to improve the transport system.

In 1922, the road from Rishon Lezion to the Jaffa-Jerusalem artery was completed and Rehovot and Nes Tsiyyona were linked to it, albeit by means of a road of lesser quality. For several years, the government delayed the construction of a road from Petah Tikva to Jaffa mainly in the hope to keep the railroad profitable, but by 1928 this connection, too, was completed. In the Sharon area, however, the transport system was undeveloped and settlement in the entire region was in its infancy.

It is noteworthy that the railroad system at this time was of only marginal utility to the citrus trade. Rehovot was connected to the system in 1920 and a spur from Rosh ha-'Ayyin to Petah Tikva was completed a year later, thus connecting the colony to the system. However, the main railroad junction was in Lydda (Lod), approximately ten kilometers inland from Jaffa and roughly the same distance from the main citrus moshavot, and it was from there that trains to Jaffa set out. The trip from Petah Tikva to Jaffa via Rosh ha-'Ayyin and Lod (including delays on the way) took about an hour and a half. Furthermore, the main difficulty for the citrus produce of Petah Tikva (and of Palestine citriculture at large) was the fact that the track terminated at the Jaffa railroad station and did not go on to the port. The port at Jaffa was connected to the railroad system by a narrow-gauge track only. It carried only two trains per day and the rolling stock was unable to haul large quantities of citrus fruits. What is more, the track itself was an impediment to the regular movement of freight and passengers to the port of Jaffa. Even this line was eventually dismantled because it did little to boost economic activity at the port. So much time was spent in waiting, unloading the fruit from the train, and loading it onto the vessels (or on the connecting train) that rail transport was unprofitable insofar as Jaffa port was concerned. In the mid-1920s, the travel distance between Petah Tikva and Jaffa was thirteen kilometers by car and forty-six by train and that between Rehovot and Tel Aviv was twenty-two kilometers and thirty kilometers, respectively. This illustrates from an additional angle the preference of haulage by road over rail transport and the concentration of most Jewish-owned citrus area around Petah Tikva. Accordingly, citriculturists at this time used the rail system mainly to ship *berara* (fruit unfit for export) to Egypt. However, the quantities of berara shipments to Egypt were not great and, by the mid-1920s, the Egyptian authorities had began to raise administrative barriers to these shipments in order to protect their own citrus industry. Not until the early 1930s, when growers began to export citrus via the port of Haifa, which was under construction, and especially after the port was officially opened in October 1933, did the railroad become an important means of transport for the Jewish citrus industry in the moshavot of Judea and the Sharon Plain.

With the onset of the British Mandate in Palestine, the number of the motor vehicles, including trucks, grew immensely and rapidly relative to the Ottoman era, and these means of transport came into initial use in the haulage of citrus fruit. However, due to the sluggish development and poor quality of the road system, camelback remained the most common conveyance until 1927. Contemporary testimonies report that convoys of hundreds if not thousands of camels, balancing full crates of citrus on their humps, reached Jaffa port every day at the peak of the shipping season. Thus, the inadequate development of the road and rail systems—occasioned by budget shortfall and the government's wish to maintain the monopoly of the railroad—continued, for the most part, to dictate the Thünenian development of Jewish and general citriculture in Palestine, namely, in concentric circles radiating from the port of Jaffa.[20]

The picture would change sharply by 1936. As Table 3.2 shows, the most significant change in the spatial distribution of Jewish citriculture was the dramatic ascent of the Sharon Plain from 5 percent of Jewish-owned citriculture area in early 1927 to 37 percent in 1936. In contrast, the share of Judea decreased perceptibly, although this region remained the principal citrus area, its orchards accounting for almost one-half of orchard area in Jewish hands at the time. Also noteworthy is the steep decline of Petah Tikva from its primacy as the center of Jewish citriculture. In 1927, the orchards of Petah Tikva accounted for 43 percent of Jewish-owned citrus area; by 1936, the share of this moshava had fallen to only 9 percent. Rehovot maintained its relative position, more or less, whereas Rishon Lezion, which in 1926 had only five hundred dunams of orchard, embarked on a rapid process of uprooting vineyards and replacing them with citrus orchards, as I have shown. Thus, by 1936 Rishon Lezion had carved out a rather respectable place among the citrus moshavot. Notably, eight thousand dunams of Jewish-owned vineyards were uprooted in 1927–1936, mainly in the moshavot of Judea, and the main purpose was to replant the area with citrus.

The change in the spatial distribution of citrus area during this time was not due to a planting moratorium in the old moshavot. In fact, these settlements increased their planted area by hundreds of percent, as Table 3.2 shows. However, the spread of citrus, as stated, transcended the old moshavot and branched into areas where citrus activity had been negligible in previous years. Overall, the data show that the "planting frenzy" gripped almost all parts of the country where citrus could be grown. Orchards were planted on the Sharon Plain and in Rehovot and the areas to its south, where the light soil was most appropriate for citrus, and were planted in the heavy soil of the inland valleys and the stony soil of Zikhron Ya'akov and its environs. One of the main reasons for this eruption of activity, of course, was the high prices that citrus fruit obtained in European markets in the 1920s and early 1930s.

The frenzy was abetted by the development of the transport system, which brought most of the relevant settlements closer to their ports of export. Here, in addition to roadbuilding and rail development, one should take account of the vigorous growth of the motor-vehicle fleet. Motor vehicles could navigate muddy roads and keep distant localities in touch even in inclement weather, as many accounts from moshavot such as Netanya and Magdiel in their early days attest.

To avoid interruptions in commerce and damage to fruit in transit, as stated, the growers needed many paved roads. The Farmers' Federation continually petitioned the Mandate Government to expand the road system and to help pay for the construction of connecting roads between orchards and main roads or railroad stations. Since the construction of secondary roads fell under the responsibility of the local councils whose areas of jurisdiction they crossed, the performance of the roadbuilding work depended mainly on the farmers' willingness to invest money in it. As for primary roads, the main problem was the government's unwillingness to complete the Jaffa-Haifa highway due to its wish to keep up the railroad's monopoly. Only in 1936, after the Arab uprising began, did the government begin to expedite the completion of this road, which by 1937 was fully paved.

The haulage of citrus freight underwent a mammoth change at this time. Jewish localities, unlike some Arab localities near Jaffa, stopped using camels to haul their crates of citrus and went over to trucks and railroad cars. Travel distances and travel time between citrus localities and the two ports of export, Haifa and Jaffa, became shorter. Citrus was shipped by train to Haifa port from the Sharon area and places even farther away, such as Rehovot and Petah Tikva, whereas localities such as Netanya and Ra'ananna, as well as Petah Tikva, were connected by paved roads (albeit not always direct ones) to Tel Aviv, Jaffa, or the nearest railroad station.

Additional factors in the relative decline of the old citriculture were the price of land, which was much higher in the old moshavot of Judea than in the sparsely populated Sharon Plain; and the urbanization process that occurred in the Judean moshavot, foremost Petah Tikva. In 1937, Petah Tikva revised its legal status from local council to city, thereby becoming the second all-Jewish city (after Tel Aviv) in the Mandate era. Other large moshavot, such as Rishon Lezion and Rehovot, obtained local-council status and underwent far-reaching urbanization.[21]

The importance of the main city of the Yishuv, Tel Aviv, deserves reemphasis in this context. During the interwar period there was a general trend (with ups and downs) of migration of Jewish farm workers to the city, where they could make a better living than citriculture offered. This, in turn, caused the share of Arab labor in the moshavot to rise steadily. Concurrently, the city became more central than ever in the activities of citriculturists and citriculture. The main agencies that imported machinery, pumps, fertilizers, and additional capital goods and services were located in Tel Aviv. The banks were also headquartered there, and even if they opened branches in the moshavot the "first Hebrew city" remained the financial hub. Therefore, relative distance from Tel Aviv affected the growers' ability to obtain information and technology services at acceptable prices and quality. The farther the growers were from the center, the more they had to pay for these factors.

THE LABOR MOVEMENT MOVES INTO CITRICULTURE

An important aspect at this time—from the late 1920s to the early 1930s—was the entry of the Zionist Labor Movement into citriculture. Until 1927, almost no orchards had been planted in nonprivate settlements. Although attempts to grow oranges in Kinneret and Degania had been made during the Ottoman era, they were short-lived and, as I have shown, citrus area in that part of the country did not increase until the onset of the "planting frenzy." In the Jezreel Valley settlements, too, few citrus orchards were planted during the early settlement era. A report to the Fifteenth Zionist Congress (1927) notes fleetingly that a small amount of citrus had been planted in Kinneret and Degania. Later reports for 1929–1933 do point to gradual development of citriculture in the Jewish Agency settlements. In 1934, Labor Movement settlements had 5,866 dunams of citrus orchards, nearly all less than five years old. By 1938, Labor-affiliated settlements had twelve thousand dunams of orchard area, mostly in the Sharon area and, especially, in the Hefer Valley, plus two thousand dunams owned by Histadrut members who lived in moshavot. By the end of the period discussed here, Labor settlements owned 13.5 percent of Jewish-held citrus area. The aforementioned change in the Labor Movement's thinking, toward affirmation of production for the market, is accurately reflected in these data.

Another aspect of the Labor Movement's entry into citriculture, noted in the previous chapter, was the establishment of Yakhin in early 1926 as a "contracting office for agricultural labor" in response to the crisis that befell the Histadrut's economic enterprises, on the one hand, and the flourishing of the private sector in agriculture, on the other hand. The purpose of Yakhin was "to enable Jewish labor to take over agricultural work in the private sector, that which exists and that which will come into being, by centralizing these activities under all-encompassing contracting arrangements." The members of Yakhin expressed the goal more clearly. The purpose in forming the society, they said, was to channel "the private capital that has begun to flow into the country so that it will create utility not only for the individual investors but also for the entire Yishuv, by creating an intersection between it and Jewish organized labor."[22]

Thus, the purpose of Yakhin's entrepreneurial activity was to serve not the individual but society at large, that is, public entrepreneurship. What is more, Yakhin was owned not by the individuals who took part in its activity—they acquired its shares only because the law required this—but by the Histadrut. If the cooperative were to be liquidated, the bylaws explained, its assets would accrue not to its shareholders but to its legal owner, the Histadrut.

Accordingly, from its formation to 1939 Yakhin engaged mainly in the planting and cultivation of new orchards, largely for private settlements but also for those of the Labor Movement. In the early 1930s, it branched into harvesting and shipping the produce of orchards that it had planted, which in the meantime had began to bear fruit, and of orchards of Labor Movement and private settlements. In 1936, when the planting of Jewish orchards came to a near-standstill, Yakhin moved intensively into contracting for picking and packing in private moshavot by establishing "contracting offices" in these localities. Yakhin employees were required to belong to the Histadrut and many were kibbutz or moshav members who were expected to settle in localities that they cultivated for Yakhin (such as the Hefer Valley).

Yakhin's working method, in planting and cultivating and in packing and shipping, resemble those of the private planting and shipping firms with

FIGURE 3.2. View of Emek Hefer with an orange grove planted by Yakhin, 1936.

which the cooperative competed. Thus, Yakhin prepared draft budgets for potential investors in Palestine and abroad, published brochures in the habitual languages of Diaspora Jews (Yiddish, German, English, French), sent representatives to various countries to promote the company, and packed and shipped fruit of producing orchards.

For the very reason of its activity in the private sector, Yakhin initially practiced a balanced-budget policy. "It was necessary to set the modus operandi of the business, especially in regard to payments," the Yakhin report for 1930 explains, "on a very healthy basis: assurance of client payments in a timely way and on fixed dates, and performance of the work in the best manner possible."[23] Unlike private companies, however, Yakhin did not treat the maximization of profit as its principal aim; instead, its chief aspiration was to attain the social goals for which it had been formed—diverting private capital that had been invested in citrus for the employment of its workers and establishing middle-class settlements by means of a Histadrut-owned entity.

Organizational and ideological affiliation with the Labor Movement was advantageous to Yakhin in some ways and disadvantageous in others. On the plus side, it gave the cooperative a comparative advantage in its interaction with investors whose consciousness or organizational platform had an especially strong national element. Yakhin also benefited from belonging to a comprehensive organizational system. Thus, Yakhin availed itself of Histadrut emissaries to promote itself abroad and of services in Palestine that the Histadrut provided its affiliates, from assistance in finance to importation of advanced technologies and help with printing and advertising.

Another important use of the Histadrut affiliation in Yakhin's economic activity was the process of vertical integration, that is, the inclusion of all business activities, from production to marketing, under one roof. Vertical integration is not always advantageous; businesses often come out ahead by purchasing a product or service rather than by producing it at greater expense. In its first decade, however, Yakhin went through a visible process of expansion that closely resembled the one that Solel Boneh underwent as it metamorphosed.

Additionally, Yakhin shared the Labor Movement's aim for hegemony. Thus, as it provided planting and cultivation services until the beginning of production, Yakhin began to perform what was known then as an "expansionary activity." Instead of merely organizing unskilled laborers; Yakhin promoted the establishment of enterprises that would provide it with various capital goods. Thus, the Eshed cooperative, formed by a group of workers from Nes Tsiyyona, manufactured and delivered cement pipes for irrigation in Yakhin's plantations, the Hasadeh group sank wells for Yakhin, and the Haargaz cooperative was established to make citrus crates out of wood that Yakhin purchased. Yakhin also integrated itself into the citrus marketing apparatus of the Histadrut.

Yakhin's annual reports show that by the beginning of the economic crisis in 1935, the cooperative had been striving to expand in many additional directions. It sent special representatives to Europe and the United States to organize groups of private settlers; engaged in land acquisition dealings in conjunction with private individuals. By 1934, the cooperative had taken possession of two thousand dunams; Yakhin even established a loan fund that lent at especially low interest (2 percent per annum) to its employees and to organizations that cooperated with it. In its annual report for 1933, Yakhin explained that its goals for the coming year would be "the formation of a new orchard and settlement company" in conjunction with the Jewish National Fund; the expansion of planting activity into the mountain areas; the establishment of contracting offices in moshavot, operating out of the Yakhin offices and under the cooperative's control; "the establishment of central packing houses in order to centralize packing at the workplaces"; and "participation in the establishment of an institute for water-supply research."

However, Yakhin's diverse plans for comprehensive activity in settlement and citriculture came to naught. The onset of the citriculture crisis in 1935 and, as the crisis worsened, the cessation of planting in 1936 and subsequent years arrested the uptrend in land preparation and planting of orchards. Furthermore, the growth of citriculture—especially in grapefruit—in the Histadrut-affiliated settlements brought a new player into the field that took a dim view of Yakhin's expansionist tendencies. Central packing houses, for example, entail central control as well. The Labor Movement settlements, which marketed their other produce via Tnuva, refused to dwell under Yakhin's roof. The two entities and their leaders embarked on a personal and organizational struggle of titans that ended with a schism among those involved in citriculture in the Histadrut system. At the end of the 1935/36 season, Tnuva Export took over the marketing of the Histadrut settlements' citrus produce, leaving Yakhin with private growers who were willing to use its marketing services. Overall, Yakhin was not active in a large number of orchards. By 1938, the cooperative planted, cultivated, and harvested about six thousand dunams of citrus, only 4 percent of Jewish-owned citrus area at the time. However, Yakhin did participate in the establishment of fourteen middle-class settlements, eight on private land and the others on JNF land. Its endeavors in citriculture are a clear manifestation of public entrepreneurship from the Labor Movement school.[24]

Returning from our discussion of Yakhin to the general development of citriculture in 1927–1936, we find that the most conspicuous trend—as noted—was the spatial expansion in the Sharon region. Several studies describe the settlement development of the Sharon area; it is not the interest of this book to do the same. It is proper, however, to reemphasize the centrality of citrus in the agriculture of the Sharon settlements. Even localities that tried dry farming first, such as Magdiel and Ra'ananna, soon shifted their investments to citriculture. These two settlements, as well as others in the Sharon area, crossed this watershed in 1926, when the great upturn in citrus planting began. Plantation villages or farms that were established after 1926, such as Netanya and Tel Mond, took it for granted that they should invest in the most profitable type of agriculture at the time, citriculture.

The most obvious example of the nexus of private initiative, citriculture, and settlement in the Sharon area was the establishment and development of *Netanya* during this time. For this reason, it is worth briefly retelling the history of Netanya from its founding in 1929 to the end of the period of concern in this study. Netanya was founded by a group of offspring of veteran farmers who sought a place to settle and farm. Most of them had grouped at the beginning of the Mandate era in an association that they named Benei Binyamin ("Sons of Benjamin"). Some settled in new established colonies such as Herzliya; others settled in existing moshavot such as Petah Tikva, Rishon Lezion, and Binyamina. Netanya was established in a totally different way. In 1929, a group of important members of Benei Binyamin—Oved Ben-Ami (secretary of the association), Gad Machnes of Petah Tikva, and Baruch Ram of Hadera, among others—formed a company called Hanotea as the settlement department of their association. In conjunction with two other societies they acquired four thousand dunams in the central Sharon region, near the seashore, and moved there in the winter of 1929.

Their purpose from the outset was to establish a citrus settlement that would be based on a combination of "absentee landlords" and second-generation farmers who would work their own orchards and those of the absentees. The settlers were hand-picked from the Benei Binyamin membership, and the initial concern was that they be wealthy enough to honor their undertakings. The absentees, who lived either abroad or in towns in Palestine, belonged to the upper crust of their Jewish communities and the country's economic elite. The entrepreneurs amassed enough initial capital, mainly by taking loans from the Benei Binyamin Cooperative Bank, to launch the venture smoothly. In planning the citrus area, Hanotea availed itself of the knowledge of local experts and of its founders, who themselves were the offspring of farmers. Before they began to plant, the entrepreneurs planned the orchard area, chose the planting and irrigation technologies that they would use, and tested the quality of the soil and the quantity of water.

Within a year, five hundred dunams of citrus were planted in Netanya. By 1932, about three years into the venture, Netanya had grown to ten thousand dunams, of which three thousand were planted with citrus. Subsequently, the orchard area expanded to 4,500 dunams in 1933 and six thousand dunams, more than any other citrus settlement in the Sharon area, in 1936.

The entrepreneurs of Netanya had especially strong relations with the authorities. Quickly they obtained a concession for the coastal lands of Netanya and established the Netanya Coast Development Company. A detailed urban development plan for the coastal area was prepared and the local council was empowered to function as a "local town-building committee," the sole zoning and development authority for the area. Most of the planned residential plots in the "development area," as the coastal zone was called, were sold off in short order. The Fifth Aliya (1930–1939), an influx of immigrants who brought capital with them, also did much to ensure the rapid growth of Netanya. Thus, within a few years it became one of the most important towns in the Sharon area, a resort on the one hand and a citriculture town on the other hand. Until 1937, when the Jaffa–Haifa road was completed, Netanya was connected with Tel Aviv mainly by the road to Tulkarm, where there was also a railroad station. By the end of that year, regular bus service to Tel Aviv and to the Hadera railroad station had been started up and the population of Netanya had grown to five thousand.

The founders of Netanya listed four main motives for the establishment of their town: (1) A *national* motive. The introductory remarks in a promotional brochure for Netanya stressed the national utility of the project as if it were self-evident: "There isn't a Jewish achievement in Palestine," the text begins, "an act of settlement, or an economic endeavor that is totally devoid of romance and pioneerism, but Netanya, the colony itself and the settlement enterprise that it symbolizes, has all these qualities sevenfold." Furthermore, in a booklet devoted almost entirely to persuading overseas investors to invest in Palestine citriculture by means of Hanotea, the concluding sentence, rendered in large print, stresses that planting oranges in Palestine helps to create Jewish national settlement and jobs for Jewish workers. (2) There was a *private economic* motive; namely, the investors wished to turn a profit or, at the very least, to safeguard the value of their money in the uncertain 1930s. According to the sources, this factor was almost as important as the national factor. (3) The *second-generation farmers* joined the project to assure themselves a place to settle. (4) The founders wished to negate the public-entrepreneurship method (coined by them as the "public-subsidy method") that the Zionist Movement institutions used and sought to maintain their social and economic independence. The founders themselves aptly summed up their underlying worldview by describing settlement in Netanya as having been

> effected under conditions based on the principles of private property, free enterprise, and personal liberty, [through which] the settlers' farms have succeeded and prospered.[25]

Confronting the planned-economy outlook or, at the least, an economic system that allows private initiative only with "strings attached," Ben-Ami and his comrades touted the socioeconomic success of their worldview. Their willingness to invest in the Netanya projects, with all the personal risk that this involved, flowed from a basic Weltanschauung that opposed the etatist-Socialist philosophy of the Zionist Movement. Both groups were Zionist in their consciousness. The founders of and investors in Netanya, however, espoused a worldview that upheld natural rights, the principle of private property, and the freedom to do what one wishes with such property. In contrast to Yakhin, which aspired to serve the public as an entity legally and conceptually separate from the individuals of which it was composed, Hanotea acted on behalf of the personal welfare of each individual who took part in its activities—the second-generation farmers who had established the town, the people who settled there, and the absentee investors. The adoption of the worldview behind Hanotea, in the estimation of those involved in the company, would result in greater welfare for both the individual and for society at large—as the success of Netanya proved.

The founders of Netanya were undoubtedly "Schumpeterian" entrepreneurs. In addition to blazing new trails, of which the establishment of Netanya was definitely a fine example, they saw opportunity—to use Kahan's turn of phrase—where others perceived only uncertainty. All sources mention the uncertainty that swathed Netanya in its early days, its distance and isolation from Jewish settlement centers, and the chronic uncertainty that accompanies such a large investment in an industry as unstable as citriculture. However, all sources also note how quickly Netanya turned the corner toward profitability and mention the town's social and economic flowering and prosperity. Contemporary sources state correctly that had the developers of Netanya not managed to obtain title to the areas near the coast and the right to plan them, the citrus crisis of the late 1930s would have dealt the town a very severe blow. Thus, it seems that two salient types of entrepreneurs joined forces in the establishment of Netanya: Schumpeterian entrepreneurs who continually broke new ground and "rank-and-file" private entrepreneurs who risked their wealth by investing it in the citriculture and "development area" of Netanya.[26]

As stated, Netanya was a good example of the influence of citriculture on the development of a section of Palestine where Jews had not settled before. However, other localities in the Sharon area also developed their citrus industries appreciably. In 1936, there were 5,100 dunams of orchards in Kfar Saba, 5,000 in Ra'ananna, and 4,500 in Herzliyya—as against only 785 dunams in the entire Sharon area nine years earlier. The moshavot also gained in population, for reasons including the transition to citriculture and the proximity of some of these localities to Tel Aviv, the hub of the Yishuv.

In Samaria, north of the Sharon Plain, Hadera had the highest proportion of orchard area (41 percent of the total), followed by Pardes Hannah (13 percent), Karkur (12 percent), and Zikhron Ya'akov (6 percent). Together, these four conspicuous examples of private settlement accounted for 72 percent of citrus area in Samaria. Other privately settled localities in the region made respectable contributions of their own.

The Galilee boosted its share of Jewish-owned citrus area from less than 1 percent in 1927 to 3 percent. In 1936 about two-thirds of this land was in Labor Movement settlements in the Jordan and Jezreel valleys, and most of it was planted with grapefruit, which was suited to the heavy soil that typified this part of the country. (See Table 3.2.)

SOCIAL AND ENTREPRENEURIAL INDICATORS IN THE LATE 1930s

Can one sketch a social and entrepreneurial profile of the private Jewish citrus growers at this time, the late 1930s? Obviously, we cannot retell the biographies of all of these personalities or even most of them, as we did for the first citriculturists in Petah Tikva and Rehovot. After all, the Jewish sector in Palestine had more than three thousand private citrus growers by the beginning of that decade. However, beyond our discussion of the activities of several of the most outstanding of them, such as the founders of Netanya, we may avail ourselves of several contemporary censuses and statistical reports to arrive at a more general characterization of those thousands of citrus growers in the late 1930s.

First, citriculturists were a minority of the population and the labor force of both the old and the new moshavot, even though most moshavot based their economies on citrus. In 1934, the Farmers' Federation conducted a census among the moshavot in order to elicit information about their economic and social development. The census showed how dependent the citrus industry was on wage labor. However, it also reinforces the claim that citriculture provided a livelihood for thousands of people, especially Jewish unskilled workers who accounted for a majority of Jewish workers in the moshavot.

Another element in the profile of these entrepreneurs is the relatively high proportion of growers, especially in the old moshavot, who lived off their farmsteads. Importantly, however, few "absentee landlords" lived outside of Palestine. Most lived in towns of the Yishuv, foremost Tel Aviv. In 1930, there were 3,364 farmers in "private-property moshavot" in citrus areas (i.e., excluding Galilee). About 47 percent of them (1,598) did not live on their farmsteads. This picture comes into sharper focus when we examine the proportion of absentees in the moshavot of Judea in comparison with other regions. It turned out that 53 percent of farmers in the old moshavot lived elsewhere whereas only 30 percent of farmstead owners in the Sharon Plain moshavot did so. In 1936, a Jewish Agency census of Jewish citriculture in the six blocs of old-citriculture orchards elicited a slightly different picture. The share of absentees in these moshavot declined from

47 percent of growers in the 1930 census to 30 percent in 1936. This seems to be another instance of the statistical imprecision that occurred frequently at the time. One does, however, get the impression that the 1936 estimate is more accurate and closer to the truth.[27]

Another notable phenomenon during this time was the accumulation of land in individuals' hands. In 1936, 58.4 percent of growers in old-citriculture localities held only 22 percent of the orchard area. The orchards at issue were small, at one to twenty dunams in area. Some 28.6 percent of growers owned orchards of twenty to fifty dunams and 13.1 percent of growers held title to nearly half (47 percent) of the citrus area in these localities. The orchards in the last-mentioned category were large—fifty dunams or more. This distribution follows a typical capitalistic pattern of capital accumulation. However, wealth did not accumulate perceptibly since most citrus areas remained in the hands of small and medium landholder-entrepreneurs.

Finally, even many growers who lived on their farmsteads practiced additional and nonagricultural occupations for their livelihood. The large investment in the orchard, the lengthy time that lapsed until production could begin, the proportion of wage labor in the industry, and the summers, during which work was not intensive, encouraged growers to augment their livelihoods with other occupations off the farm.[28]

SPATIAL DISTRIBUTION IN ARAB CITRICULTURE

Although the sources on Arab citriculture are quite scanty, below I attempt to profile the spatial distribution of Arab citriculture on the basis of the sparse data available. The evidence suggests that Arab citriculture, like Jewish citriculture, was distributed largely in accordance with distance from the market. As I showed in regard to Jewish citriculture, most of the coastal plain between Zikhron Ya'akov and Gaza had a soil structure that was generally suitable for citrus. At the end of the Ottoman era, most Arab-owned orchards were in the vicinity of Jaffa. In 1935, too, most Arab citriculture took place in that area, although, as in the case of Petah Tikva, the Jaffa vicinity had begun to lose its exclusivity. Another sizable share of Arab-owned citrus orchards was planted in the vicinity of Ramle, which paralleled the bloc of moshavot comprised of Rishon Lezion, Nes Tsiyyona, and Rehovot. Insofar as the sources speak of the matter, they indicate that Arab citriculture, like its Jewish counterpart but to an even greater extent, was an enterprise made up of private investors. Its development, too, seems to have followed a general spatial trend: the Jaffa area first, followed by a thrust in the direction of Ramle and its environs, and later northward to the vicinity of Tulkarm, alongside the Jewish-held Sharon Plain.

By 1918, Tulkarm had a railroad station and was connected to Jaffa by an asphalt road and a hard-packed dirt road. According to the sources, the

TABLE 3.3
Geographical Centers of Arab Citriculture, 1935

District	*Dunams*	*Percent*
Jaffa	43,844	38.8
Ramle	30,228	26.7
Gaza	17,404	15.4
Tulkarm	12,291	10.8
Acre	9,014	8.0
Others	290	0.3
Total	113,071	100.0

Source: Issa Mustafa Alami, *Some Aspects of the Development of the Palestinian Peasant Economy and Society, 1920–1939* (PhD Dissertation, Edinburgh, Edinburgh University, 1984), 471.

city became the hub of a rather large citrus industry that accounted for 11 percent of Arab citriculture in 1935. The industry was based on the capital of affluent Arabs from Jerusalem and Nablus, as well as local investors, some of whom had accrued their wealth by selling land to Jews. Citriculture also developed in the Gaza and Acre subdistricts, which had marine ports and high groundwater tables. In the latter area, too, the government had established an agriculture research station that dealt with citrus, among other things.

Summing up the chapter, we may say that the spatial distribution of Jewish and Arab citriculture was strongly influenced by distance from ports of export. The more undeveloped the transportation infrastructure was, the more the spatial distribution resembled von Thünen's model of concentric circles. As a modern transport system took shape, the concentric circles gave way to a complex structure that included the transport alternatives and their costs. The modernization of transport infrastructure at this time, which was prompted partly by citriculture and which nourished it, made a very significant contribution to the spatial expansion of transport and, perforce, of citriculture in Palestine.

As for the profile of the private citrus entrepreneurs, a diverse picture has come to light. Some of the individuals in question were "Schumpeterian" entrepreneurs who broke new ground and changed the social and economic status quo. The large majority of growers, however, were private entrepreneurs who took actions at considerable risk and under conditions of uncertainty. This was especially so in citriculture, which depended on the international market and in which investments took long to mature. Thus, one should not be surprised to discover that many citriculturists did not live on

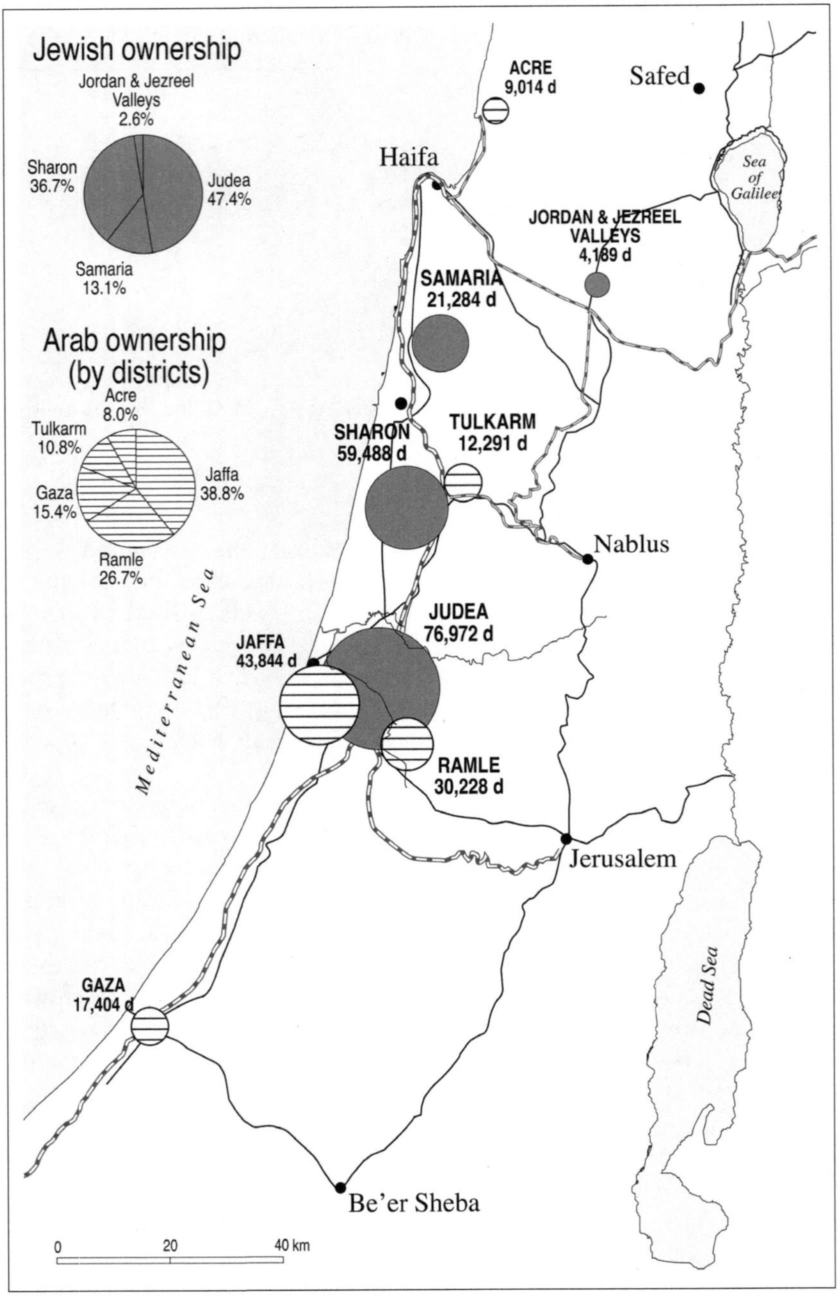

Map 3.2. Spatial distribution of Jewish and Arab citricultures, 1935–1936.

their farmsteads, instead dwelling in towns and presumably pursuing additional sources of livelihood. Even in the moshavot themselves, quite a few growers and members of their families sought to make a living by taking on additional jobs, in hopes of augmenting their citrus income—which, as the number of young orchards increased toward the end of the period researched, did not materialize at the level expected.

PART III

Technological Innovations

FOUR

From Jaffa to Petah Tikva

Technological Development in Citriculture during the Ottoman Period

DEFINITIONS AND THE STATE OF RESEARCH

TECHNOLOGICAL INNOVATION is a proven and important factor in economic growth at large and, in particular—for better and worse—in the economic growth and dominance of the West in modern society.[1] Furthermore, both the experience of economic development in Western countries and the attempts to induce similar economic growth in third world countries, insofar as they have succeeded—point to the vast importance of economic and technological development in agriculture in the overall growth of any economy. This is not only due to the migration of labor from village to city and its consequence, the creation of incentives for further technological innovation in agriculture and for more intensive research and development; but also due to the expansion of the urban market (fueled by migration from rural areas, among other factors) for farm produce, an expansion that also creates incentives for technological innovation in agriculture. These processes also create incentives to invest in infrastructure, develop human capital, and expand the field in which agriculture, manufacturing, and services intersect.[2]

A comprehensive examination of each of these factors (R&D, production, infrastructure, education, etc.) and its effect on the technological development and growth processes of the economy of Palestine during the period reviewed surpasses the limits this discussion. Neither can this book subject the interrelations between these factors and the citriculture industry to comprehensive examination. Several of these fields were discussed above, and technological innovation during the Mandate period and its contribution to overall economic growth have been described in several detailed macroeconomic studies by Jacob Metzer, among others.[3] Thus, we will treat these elements as a general frame of reference only. Their intereffects with the citriculture industry (in respect to transport infrastructure or citrus R&D, for example) will be noted where possible.

In any event, the main topic in this part of the book will be the development of technology in Palestine citriculture, especially its Jewish sector. The main question—if we may adopt a term used by one of the most prominent researchers of technological development, Nathan Rosenberg—is what happened inside the "black box" of technological development in citriculture. Following Rosenberg and additional studies on economic history, the technological development of Jewish citriculture in Palestine will be analyzed from a combined historical and economic perspective, on the basis of quantitative and qualitative data.[4] Few studies have discussed specifically the history of technological development in the Jewish community of Palestine in the late Ottoman era and the Mandate years. I do not intend to discuss them all.[5]

In research on technological innovation, a conventional distinction is drawn between technological *invention*, a radical act that creates a new and previously nonexistent situation, and technological *innovation*, which improves existing technologies. Some scholars treated technological invention and technological innovation as wholly unrelated and assigned the entire influence of technological change to the revolutionary breakthrough of the "grand invention" and the person or party behind it. Others, however, underscored the importance of technological innovation and the continual improvements of the existing or new technologies; while some scholars, such as Vernon Ruttan, recommend the obfuscation of all differences between invention, a one-go act, and continual improvement of existing technologies, that is, technological innovation.[6]

However, I believe it proper to accept Joel Mokyr's reasoned differentiation, which argues that each category is important in itself and that the two should be considered complementary. Mokyr defines the grand technological invention (or *the macroinvention*, as he terms it) as the ab nihilo creation of a radical new and utterly unprecedented idea, and technological innovation (*the microinvention* in his terminology) as the sum of many small steps that improve existing techniques, enhance economic efficiency, and facilitate the transformation of macroinventions into useful and cost-effective techniques. Without the macroinvention, which transcends the accepted rules of economic explanation, the innovative process cannot be sustained. Thus, grand invention and nonheroic, gradual innovation are complementary processes.[7]

Two additional issues pertain to the issue at hand: (1) the distinction between the *diffusion* of a technological innovation and its *adoption*; (2) the factors that affect the diffusion of a technological innovation and the decision to adopt it. In regard to the first issue, research accepts the assumption that a technology is adopted at the individual-firm level whereas an innovation is diffused over a period of time and amidst a population (of people or firms). Many studies that deal with this issue, however, do not apply this distinction because for a socially and economically significant process of technological spread across firms or individuals to take place, adoption and diffusion must be interdependent, that is, in the absence of one process the other does not take place. This is especially important in the type of agriculture discussed here—one that is composed of hundreds of small firms.

As for the second issue, many factors affect the decision to adopt or reject a technological innovation: general factors related to the structure and institutions of society, society's basic value system, the population's levels of schooling and economic and social modernization, and considerations that are pronouncedly economic. In regard to the last-mentioned—the factors on which this study focuses—the literature shows that although there is no consensus about any particular model that explains the diffusion or adoption of technological innovations in agriculture, most scholars agree about several salient economic components. These include the levels of supply of and demand for technological innovations; the expected profit due to the adoption of a technological innovation; the size of the firm; price ratios among the basic factor inputs (land, labor, and capital); the ratio of these prices to the cost of investing in and using the technological innovation; and the pace of adoption.[8] The discussion to follow will examine the case of Jewish citriculture in Palestine in view of the research literature and will examine the singularity of this industry vis-à-vis the conventional wisdom.

TECHNOLOGY IN THE TRADITIONAL CITRUS ORCHARD

Several sources from the Ottoman period provide an accurate picture of the Arab citrus orchard before Jewish growers introduced modern methods. The most important source is Aharon Leib Fellman's aforementioned *Ma'ayan Ganim: Rules for Planting Gardens and Citrus Orchards in the Holy Land* (1891) (Hebrew). In his preface, Fellman informs the reader that he has "nine years' [experience] in gardening and citrus growing in the Jaffa area." The result is a firsthand account of local methods and practices of cultivation at the dawn of Jewish citriculture. Other works that complement Fellman's detailed account include *Divrei Mordechai* (The Words of Mordechai) (1889) (Hebrew) by Mordechai Diskin; an 1893 report by a government engineer George Franghia on the cultivation and irrigation of oranges in the Jaffa region; and a comprehensive article by the Zionist agricultural experts Aaron Aaronsohn and Selig Soskin in 1902, "Die Orangengärten von Jaffa." The last-mentioned work reviews the commercial and marketing methods of Jaffa oranges and compares the commonly used "Arab methods of cultivation" with the latest "European methods" that Arab, Jewish, and European growers in Palestine were already adopting.[9]

Fellman's goal, as the preface explains, is to produce a useful manual for middle-class Jewish investors in Palestine and elsewhere who wish to base their economic activity on agriculture in Palestine. As Fellman explains it, "This book will guide and instruct settlers in this country and our brethren abroad . . . by giving them correct knowledge in the matter on which all Jews' eyes are pealed" (the new settlement enterprise in Palestine). The preface also notes—as did others at the time—that cereal agriculture in Palestine is not profitable and that "[i]n view of the current condition of our country, it is much more advantageous to own an orchard than a productive field." The entrance barrier in citriculture, Fellman states, is the large capital investment that acquiring a productive orchard entails. For a smaller initial investment, one may buy some land, prepare it for planting, and cultivate it until it begins to produce in reasonable quantities. However, one must also invest in agronomic know-how and assure oneself a sufficient livelihood for years, until the investment begins to provide a return. For these reasons, Fellman claims Jews who lack ample capital resources refrain from investing in citriculture and the well-endowed (who acquire producing orchards) fall victim to slick-tongued traders who sell them parcels at prices that are inflated "two or four times over." Fellman's solution to this dilemma is to show small investors that, on the basis of rather simple know-how (which his book will present), they too can invest in the establishment of new citrus plantations and obtain a good return on their investment.

Fellman's book is divided into thirteen short chapters that explain various aspects of preparing and cultivating an orchard up to the end of the sixth year. In the seventh year after planting, Fellman states, the trees will produce fruit and "reward their owners handsomely." Aaronsohn and Soskin are more pessimistic about the maturity of the investment (about seven years after planting); Franghia suggests that it would take seven to eight years. Be this as it may, Fellman devoted his last chapter to a detailed financial calculation of the process of bringing the young orchard to fruition, from the purchase of land to the first viable harvest in the seventh year. According to Fellman's calculus, the investment in a thirty-six-dunam orchard, from purchase of land up to the first harvest, is about ten thousand francs.

Franghia, whose reckonings pertain to the 1890–1891 period—the time when Fellman published his book—gives much higher estimates. For example, a dunam of land before preparation and planting, he says, costs between 91 and 106 francs. Fellman estimated this at only fifteen francs. However, Franghia's data leave much to be desired; Fellman's computations are much more detailed and leave no doubt about the author's thorough familiarity with the topic. Neither Fellman nor Franghia show how much revenue the orchard would generate after it begins to produce exportable fruit, except for

Fellman's assertion that such orchards would "reward their owners handsomely" and Franghia's claim that the profitability of the industry had doubled between 1880 and 1890. For this reason, we cannot calculate the profitability of either author's orchard. However, the testimonies of Franghia, Fellman, and others about the steady uptrend in citrus exports and the prices of suitable land show that citriculture was a profitable endeavor that was becoming increasingly so over the years.

The Jewish experts at the time tended to relate to Arab citriculture—and a fortiori the Arab subsistence farm—as a "primitive" nativist phenomenon that operated without economic logic and was characterized by laziness, inefficiency, technological and mental backwardness, and an irrational deterministic view of economic activity (and of human activity at large). This perception of traditional agriculture was not only asserted by Jewish settlers but was (and in some cases still is) accepted among many European experts who viewed non-European societies through a modernistic, Eurocentric prism.

In the early 1960s, however, the American economist Theodore Schultz (and many of his successors) showed that traditional agriculture operated rationally, attained a high degree of economic efficiency, allocated its resources effectively, and operated on the basis of well-balanced economic reasoning. Traditional farmers were unwilling to invest more labor and capital in their farms, Schultz showed, because the marginal rate of return was too small to give them an incentive to do so.[10] Arab citriculture in the early 1890s, of course, was not the sort of traditional farming industry that Schultz described; after all, it produced for the international market. However, as the Jewish experts discovered and as contemporary documentation corroborates, as I show below, local Arab citriculture in Palestine of the early 1890s had pronouncedly traditional characteristics, such as intensivity of labor and scarcity of capital. Below we ask whether the principles of which Schultz spoke, foremost the principle of efficiency in resource allocation and the economic rationality of citriculture activity, may be applicable to the local Arab citriculture of late-nineteenth-century Palestine.

We now examine the technological components that were conventional in local Arab citriculture before the introduction of significant improvements and we probe their underlying economic logic. To accomplish this, we discuss five main factors in the establishment of young orchards, from preparation and planting up to the production of an exportable crop: (1) choice of location; (2) the well and the irrigation system; (3) readying the area for planting; (4) planting distances; and (5) propagation.

As noted at the beginning of this study, two principal factors determined the *choice of location* of a citrus orchard in Palestine in the late nineteenth century: availability of groundwater and proximity to a port. Above, we discussed the proximity issue and the spatial expansion of citrus areas. Here we examine the technological aspect of the orchard location, namely, the availability of groundwater. This factor is important because the trees had to be irrigated regularly six to seven months of the year. A citrus tree needs large quantities of water throughout the year to sustain its major processes, such as the adsorption of minerals from the soil and of gases from the atmosphere, and the formation and transfer of carbohydrates. Rainfall sufficed during the rainy months of an ordinary year, but in the seven dry months (May-November) most orchards in Palestine (except for the few that were located near streams) were irrigated from wells. This explains the Arabic term for an orchard—*beyara*, a cognate of the Hebrew *be'er* (*bir* in Arabic), denoting the well that was situated in every grove at that time.

Since local growers had not begun to mechanize their wells, the availability of groundwater within ten to twelve meters of the surface was a necessity for regular and cost-effective operation. Accordingly, most late-nineteenth-century citriculture in Palestine took place on the coastal plain, mainly in the vicinity of Jaffa, which was not only close to a port but also had the highest water table. "Beyara," Fellman remarks, "are planted only in the lowlands, where . . . it is easy to find water near the surface, rather than a place of high elevation where a very deep well must be dug at steep expense and where mules and horses can hardly bring up the water."[11]

Wells and irrigation systems were installed in the traditional way. The system included a well, a pumping facility of the waterwheel type, an irrigation pond, ditches to deliver the water, and irrigation basins around the trunks of the trees. All these actions were performed either manually (excavating the well and the pond, building the pumping facility, digging the ditches) or by means of beasts of labor (pumping the water).

The well and the pond adjacent to it were excavated in the upper part of the orchard so that water could flow to the trees continually. The diameter of the well was made large enough to install the waterwheel. The waterwheel (*nuriyya* in Arabic), the traditional Palestine pumping facility, was made of two wooden wheels, one horizontal and one vertical. The beast of labor (usually a mule or a camel, but sometimes a horse) turned the horizontal wheel, which spun the vertical wheel by means of a set of cogs. Connected to the vertical wheel was a chain of vessels that raised the water from the well and channeled it to the pond. Originally the vessels were of earthenware; they were subsequently replaced by wooden boxes that, at a later phase, were lined with metal. The pond, made of stone and plastered to prevent seepage, had various functions: to keep the flow of water smooth, to impound water for future irrigation, and to slightly heat the chilly water of the well. The ditches, also lined with plaster for sealing purposes, crisscrossed the orchard and were connected to the pond. Basins were dug around the trees and had an opening through which water was channeled from the ditches. All activities related to installing the pumping system and the water-distribution system,

FIGURE 4.1. Traditional water pumping system.

from digging the well and the pond up to the act of irrigation itself, were performed by means of manual labor.

The output of the traditional water system was limited by the output of the waterwheel, which came to between five and eight cubic meters per hour only. The traditional pumping facility also limited the depth of pumping to twelve meters. Every meter in the chain of vessels, when totally filled, weighed about seventy kilograms. Thus, the animals that spun the waterwheel lacked the strength to hoist water from a depth of more than ten to twelve meters. This made it very hard to plant additional orchards, since the quantity of easily accessible groundwater on the coastal plain was rather limited. According to Fellman, a flow of five to eight cubic meters per hour sufficed only for twenty-seven dunams of orchard, "and one who adds dunams digs two or three wells, commensurate with the size of the orchard." Several growers had found an alternative: purchasing a double waterwheel mechanism and hitching it to a team of animals. By so doing, one could extend the depth of pumping (at the same flow) to between twelve and fourteen meters. Fellman did not offer his readers this option, evidently because it was too expensive for the medium-level investors to whom Fellman addressed his writings. Be this as it may, the waterwheel did not produce enough water to irrigate the entire orchard in one go; irrigation had to be performed throughout the week, one-sixth of the orchard each day. Accordingly, the depth-of-pumping limitation constrained the expansion of citriculture beyond areas where groundwater was close to the surface and the limitation of the paltry output of the waterwheel constrained the expansion of individual orchards. These constraints were not surmounted until new technology was introduced in the water pumping and delivery systems.[12]

Preparing the land before planting was also a manual enterprise. It was essential to rid the parcel of weeds (especially couch-grass) before planting because they competed with the trees for nutrients and water. Otherwise, the weeding usually had to be done at greater expense after planting. In local agriculture, the entire orchard was hoed by hand. Additionally, special laborers were hired to comb the orchard area after each winter, weed it, and hoe the land again. Another phase in readying the land, evidently not invoked until around 1900, was hoeing to a depth of between sixty and one hundred centimeters wherever planting was intended. This depth hoeing, known as *bahar*—the term is regularly cited in Zionist contemporary accounts as one of the toughest challenges that unexperienced Jewish workers faced—resulted not only in systematic weeding of the orchard area but also in loosening the soil and readying it for planting. Notably, in Fellman's reckoning, weeding before planting was the second-largest nonrecurrent expenditure in the budget of a fledgling orchard, exceeded only by the expense of installing the water pumping and distribution system.[13]

Planting distances was one of the most important issues in orchard technology during the period at issue; it was also a source of much disagreement among experts. In the late nineteenth century, local orchards in Palestine were planted in severe density, usually at intervals of 1.5–2 meters between trees and 3.5 meters at the most—more than two hundred trees per dunam, according to Aaronsohn and Soskin. Since such congested conditions ruled out the possibility of planting trees in straight rows, Arab orchards had an "unaesthetic" appearance that Western observers inevitably linked with "primitivism" and technological and mental backwardness.[14]

Propagation of citrus varieties, especially in regard to the Shamouti, underwent many metamorphoses during the period at issue. At first, Jewish citriculture adopted local methods in this respect, too. Citrus propagation was divided into two main phases: propagation and growing of rootstock and grafting of a scion on the rootstock. The rootstock used in traditional citriculture in Palestine was the sweet lemon. The local grower designated a special area in his orchard as a nursery for the propagation of rootstocks. The propagation was done by means of shoots that were set aside for this purpose, placed in the nursery soil at very small intervals, and left there for two years. During this time, the shoots grew and became saplings with crowns and roots. At the end of the two years, the grower moved the sweet-lemon saplings to the orchard and planted them in pits that he had prepared. The spacing was also determined at this time. The rootstock saplings grew there for another two years before the chosen species or variety of citrus was grafted onto it. The most common variety, as stated, was the Shamouti. Thus, four years lapsed from the beginning of propagation of the rootstock until the grafting (two years in the nursery and two years in the orchard). Afterward, another two years or so passed until the tree began to bear fruit—six full years after the shoot was placed in the soil.[15]

Another important point is that, in contrast to views expressed by Jewish farmers, growers, and even some researchers, the sources in our discussion leave no doubt that the local Arab citriculture used an input that is considered modern, animal manure. Local citrus growers did this in the belief that manure had healthful and growth-stimulating qualities. If so, why did Fellman not recommend greater use of manure, as Jewish growers chose to do later? The sources do not provide a clear answer, but both Franghia in his budget and Aaronsohn and Soskin in their remarks suggest that manure was a rather expensive input. In Arab agriculture of the time, the cattle and sheep that produced the manure grazed in large areas of land and were not kept in corrals. This made the manure rather expensive to gather and transport. Furthermore, the manure that piled up in the villages was of poor quality, devalued by the presence of soil and other substances. Moreover, it was also used as a fuel after it was dried, and this alternative use affected its price. Finally, the density of planting and the assumption, adduced from Fellman's writings, that the manure should be placed in the basins where the trees are planted and not distributed across the entire orchard, show that fertilizing expenses climbed in direct proportion with trees per unit of land.[16]

ECONOMIC EQUILIBRIUM IN LOCAL ARAB CITRICULTURE

In the introduction of innovative technology in agriculture, a distinction is made between two main types: "labor-saving" mechanization that lowers the cost of labor relative to the cost of land and, especially, of capital; and "land-saving" biological and chemical technology that brings down the cost of land per unit of output relative to the costs of labor and capital. Examples of mechanical technology in the traditional orchard might be the preparing of land by means of machinery operated by animal or motor power or greater efficiency in the irrigation system. Biological technology is usually related to varieties that are more productive than local ones. A chemical technology normally has to do with fertilization that increases production per land unit. Thus far, our account of the cultivation methods practiced in the local Arab orchard shows that the methods were nearly always labor-intensive, not capital-intensive. Neither mechanical nor chemical technology was well developed. The mechanical technologies that were practiced including preparing of the land, a manually dug well, a wooden pumping facility operated by animal power, and distribution of water by gravitation. Chemical technology, that is, fertilization, was practiced on a very small scale. Do these methods comply with economic logic? One may answer the question largely in accordance with Schultz's outlines. Following Schultz, one may presume that the local Arab orchard was in economic equilibrium as an economic unit. One may also presume—unlike Schultz—that the technological know-how of local growers in the early 1890s was in flux; they, too, were exposed to new technologies. Importantly, most local growers were investors who sold their produce in the international market. In the second half of the nineteenth century, as the research literature has shown, a group of local entrepreneurs and merchants formed in Palestine and contributed much to the country's economic and technological development; a few of them also engaged in citriculture. Local growers could familiarize themselves with the new technological knowledge by visiting the German-Templer colonies, studying the Jewish moshavot (which by then been benefiting from the patronage of Baron de Rothschild and his experts' knowledge for quite a few years), and additional sources (e.g., relations with foreign consuls or experts such as the government engineer, Franghia). Indeed, as Aaronsohn and Soskin note several times in their article, affluent Arab growers also introduced technical enhancements in their orchards.[17] However, as demonstrated above, few local growers did so to an appreciable extent; most activity in the orchard was labor-intensive. How can one explain this?

The main constraint, it seems, was the low cost of labor, due to its abundant supply, as against the high cost of the capital needed to introduce technological innovations in citriculture. Scanty availability of the requisite technology, the need to invest in know-how in order to operate it, and the shortage of maintenance and repair services combined to make capital much more expensive than labor. The cost of investing in the new technology relative to the price obtained for citrus fruit in foreign markets was another disincentive to the introduction of new technology in traditional citriculture. Furthermore, many growers did not have enough capital of their own to invest in new technology, and any attempt to make up the shortfall would have made the investment even costlier.

The most severe technological limitation seems to have been water technology. A wind-powered pumping system that provided homes with running water had been installed at the Templer colony of Sharona back in 1879. In the 1880s, Rothschild helped the settlers of Rishon Lezion to perform deep drilling for water, and in 1890 two steam-powered water pumps were installed at the wells of the wineries of Rishon Lezion and Zikhron Ya'akov. However, these systems were expensive and the investment capital for them came from parties outside the local economic system.[18] By inference, local growers could not afford to invest in such systems at that time. Hence, as stated, the specific pumping technology limited the maximum orchard area to about thirty-six dunams. It was the grower's task to maximize production per land unit in terms of the factor inputs that he employed in his orchard. This probably explains the density of the traditional orchards.

Using this information as our point of departure, we may explain matters in two ways. First, we may presume that at their level of agronomic and economic knowledge, the local growers did not realize that wider spacing might result in a larger harvest per land unit. Accordingly, in their estimation, the more trees, the better. This resulted in a diminishing marginal output per dunam relative to a smaller number of trees.

The second explanation presumes that the local growers did understand the advantage of planting fewer trees per dunam. To compensate for fewer trees, however, they would have to invest more in land-saving technology, that is, fertilizer. This entailed additional capital, since fertilizer was a rather expensive input. They had no source for such capital. If one may judge by the number of cases exported from Jaffa Port during the 1890s, the demand for oranges in the export markets was constant in 1890–1900. The economy of Palestine, including its citrus industry, did not begin its economic "great leap forward" until after 1900, as Nachum Gross (and others in his wake) showed. By inference, there was little capital formation among local Arab growers at that time. Accordingly, given the alternatives of investing more in fertilizer and investing in dense planting—and in consideration of the fact that the labor cost had not changed—they continued to plant at close intervals even though they were familiar with the modern technologies. What is more, trees that are so closely spaced neither develop wide crowns nor grow to significant heights. In contrast, a widely spaced orchard exposes the trees to more sunlight and gives the branches more room to grow. Accordingly, the trees attain greater height, breadth, and yield. Harvesting such an orchard entails an investment in capital goods (special ladders, fruit satchels, protective clothing, etc.). As I show below, citriculture in California was practiced exactly this way, with capital intensivity at all phases of production. In Palestine, however, labor was much more abundant than land and capital. Indeed, women and children took part in the harvest in Arab (and in some Jewish) orchards not only in the late nineteenth century but also during the Mandate era. Their wages were half those of the men. It is obviously more efficient, in economic terms, to use women and children this way if the trees are small. Male pickers, too, are more productive if they do not need to climb ladders as they work. Isaac Rokach, one of the most prominent personalities in Jewish citriculture during the Mandate era and one who had known the industry inside and out since childhood, gave a picturesque account of these matters in his memoirs. Arab contract laborers, Rokach noted, would sing, "We want little trees and huge fruit" as they harvested.[19] Thus, it seems that even Arab citriculture in the late nineteenth century acted rationally and had the clear goal of allocating available capital and labor resources with a maximum of efficiency.

AN ACCELERATED PROCESS OF MODERNIZATION

A change in the equilibrium in local Arab citriculture by expanding the cultivated area of orchards, by increasing the output per dunam, or by doing both, could be achieved by an increase of capital formation among local growers or by import of capital. As stated, the bottleneck in local citriculture was water technology. In 1891, Fellman wrote that "almost no empty space remains" in the Jaffa area for orchards. Thus, since it was not cost-efficient to sink deep wells in areas far from the coast due to the state of technology, "[t]hey now say that orchards should be planted in other towns along the coast, such as Acre and Haifa." The constraint of pumping technology as against the demand for citrus also lurked in the background of Franghia aforementioned plan to increase the planted area by using the waters of the Yarkon River. Indeed, the high-quality oranges of Palestine were in increasing demand over the years, especially in the most important fruit market of the time, Britain (where additional types of fruit, especially apples and bananas, were also in rising demand). This was reported both by Franghia in 1893 and by Aaronsohn and Soskin in 1902.

Thus, brisk demand for citrus fruit in the generation preceding World War I prompted growers to plant more orchards and export more cases. Citrus area expanded from 7,500 dunams in 1890 to 12,000 in 1904 and 30,000 on the eve of World War I. The increase in cases exported was also impressive: from 260,000 in 1895 to 500,000 ten years later and 1.55 million in the 1913/14 season.[20]

How was the increase in planted area accomplished? In the main, it happened because the obstacle of pumping groundwater from depths greater than twelve meters was overcome and the flow of the traditional well was boosted. First, the transmission system and water vessels of the traditional wooden *nuriyya* were reinforced with iron. Second, internal combustion motors were installed to run them. The motors, first introduced in early 1897, burned kerosene—which had been used in Palestine since the late 1860s—or coal. Now water could be pumped from depths of up to twenty or thirty meters and the flow increased from between five and eight cubic meters per hour to between twenty and forty cubic meters per hour. Thus, a grower who could irrigate a twenty-seven-dunam orchard at a flow of between five and eight cubic meters per hour could irrigate 110–135 dunams at a flow of twenty to forty cubic meters per hour. With the increased flow, individual orchards could expand and plantation areas could spread into areas where the water table was deeper. By early 1898, a year after the first internal combustion motors were placed in service, the number of engines in the orchards of Jaffa had climbed to about sixty.[21]

Thus, expectations of an income flow from a citriculture investment, prompted by the steady increase in demand for Palestine citrus, created demand for new technology that would solve the problem of pumping depth and water flow. By 1906, about 220 engines—many of them produced by the Wagner brothers of Jaffa (agents of the Deutz enterprise, who also provided growers with maintenance)—were toiling in the orchards of Palestine. The Leon Stein workshop in Jaffa, agents of Hornsby Motors, sold about two hundred engines between 1905 and its last year in business, 1910. (Not all of them were sold to citrus growers; some were used to run flour mills.) The pace of sales and the fact that enterprises other than those of Stein and Wagner were set up during this time to sell and maintain the devices show that this was a profitable and economically rational business. (Wagner made a fortune; Stein went out of business in 1910 due to poor management.)

Details about the curve of the diffusion of the internal combustion motors are scanty, especially in regard to orchards planted on the basis of the traditional cultivation methods. The fragmentary data indicate that at least with regard to the Jewish colonies, which were far from the coast and where the water table was more than twelve meters down, the adoption and diffusion of motors and the pumps that they powered followed a bell curve. This corresponds to the accepted research perspective on the adoption and diffusion of technological innovations, which describes this process as occurring at the pace and over a time that result in a bell curve or an S-curve.[22] Presumably, too, in areas near Jaffa, where the groundwater table was higher than in the Jewish colonies and the nuriyya had long been in use, local growers were less inclined to adopt the motor and the pump. Thus, adoption was evidently incomplete; some spurned it for economic reasons. However, for growers in the Jewish colonies, which were far from Jaffa and its convenient water table, the local option of pumping by means of the nuriyya hardly existed. Furthermore, the private entrepreneurs who invested in Jewish citriculture at its outset had an advantage that many local growers lacked—enough capital to invest in a pumping system.

As noted earlier in this study, the introduction of the internal combustion motor and the pump in Palestine citriculture complied with several main rules in the adoption and diffusion of technological innovations: expectation of a gain by replacing the old technology with the new; the importance of the supply side in diffusion—in our case, Wagner, Stein, and other agents of motor and pump manufacturers who diffused the new products among the growers—the city (Jaffa) as a center of information and technology services for outlying settlements; and the importance of scale economies for the agents of innovation (here, because sales were brisk relative to the number of orchards). The adjustment of the wooden nuriyya to the need to run faster, due to the introduction of the mechanical motor, shows that a process of "learning by doing and using" was taking place. Nathan Rosenberg and others cite this process and others like it as important factors in the diffusion of technological innovations because they tailor the innovation to local conditions and reduce its installation and operating costs.

Insofar as is known, these changes in pumping technology were introduced in the citriculture of Palestine, especially in the Jewish sector, irrespective of the actions of the main agent of technological change in Palestine in the late nineteenth and early twentieth centuries, Baron Edmond de Rothschild's administration. What is more, the changes were diffused by Jewish and non-Jewish agents to the Palestine citrus industry at large and not among Jewish settlers only. In other aspects of local citriculture, however, Jewish growers did the most to enhance the technological efficiency of the orchard. The enhancements were facilitated, as several researchers have shown (and as noted above), largely by Baron de Rothschild's investments in human capital, capital goods, and know-how, which created conditions that abetted the expansion of citrus plantations to areas outside the vicinity of Jaffa.

A salient phenomenon in the movement away from the local citriculture model was the increase in orchard size. In 1902, according to Aaronsohn and Soskin, the conventional local orchard area was twenty-four dunams. Fellman corroborates this by reporting a similar area, twenty-seven dunams. Where modern methods were used, Aaronsohn and Soskin reported a size twice as large, forty-nine dunams.

It became possible, and in large areas necessary, to increase orchard size by replacing manual preparation and deep-hoeing of the soil *(bahar)* with deep ploughing. This was done by means of the Göpelpflug, as it was known in German, or the French term *manège*, as it was better known in the colonies. The manège was a power system that was operated (for various purposes) by several teams of horses that spun a wheel to which a drum was attached. A metal cable was wound around the drum and was connected, in this case, to a plough. As the horses spun the manège, the plough was dragged and the ploughing action took place. The depth of ploughing was forty to eighty centimeters and it required two to nine teams of horses. According to Eliahu Scheid, Baron de Rothschild's chief supervisor of operations in the Jewish settlements in Palestine, one day of ploughing by manège saved about one hundred days of labor using the bahar method. A mechanical substitute for the manège was the "locomobile," a portable steam engine to which the metal cable was attached. Notably, both the manège and the locomobile were used at the time for several additional activities that required towing power. According to Aaronsohn and Soskin's cost calculations, the only activity in which the price per unit of land was lower in "European" orchard cultivation than in "Arab" cultivation was soil preparation by means of manége relative to the bahar method.

As orchards became larger, tree spacing became more generous. The growers believed that wider spacing would make each tree more productive by exposing its leaves to more sunshine and giving its branches more room to spread and grow. Aaronsohn and Soskin report that the tendency among modern growers was to plant at intervals of 3.5 to 4 meters. These intervals, like the other changes in the orchard, made "European" orchards look much different from local orchards; the former, with their straight rows and spacious planting, had a "modern" appearance.[23]

The first-year investment was almost twice as large in an orchard that used modern methods as in one prepared and planted in traditional ways. The increase in costs was evident in almost all respects except for excavation of the well, which was no costlier than before, and soil preparation, which was 37 percent cheaper using the "European" method (the manège) than the traditional method (bahar). The cost reduction occurred because growers rarely purchased a manège; instead, they rented it from the German Templers or others, such as the Baron's administration, which had been importing mules, manège ploughs, and locomobile ploughs from France since the 1880s.[24] This absolved farmers from having to invest in importing and maintaining this instrument; what is more, the size of the orchard mitigated labor expenses (i.e., allowed scale economies to exist). In any case, the greatest increase in expenses was incurred in replacing the nuriyya and its paraphernalia with an internal combustion motor and a pump. Aaronsohn and Soskin did not provide a detailed reckoning of revenue and expenses, least of all for the "European" orchard; they merely noted that the introduction of the new innovations would enhance profitability relative to the traditional orchard. Thus, it is hard to determine on the basis of their data whether their claims were well founded. It is clear, however, that by 1914 most Jewish citriculture in Palestine more closely emulated Aaronsohn and Soskin's "European" model than the traditional one. Nawratzki's comparison (1911) of types of food agriculture in the Jewish sector declares citriculture, in its modern incarnation, the leader in net return per land unit.

Table 4.1 shows clearly that citriculture, while the least profitable of the three orchard crops surveyed, had the largest annual turnover (net revenue per dunam) at 6,500 francs per fifty-four Turkish dunam orchard, a size that was presumably typical of Jewish citriculture at the time. The average vineyard, as we may deduce from the sources, was thirty-eight dunams. Thus, the annual net revenue per average vineyard was 760 francs. For comparison purposes, the average wage of a Jewish manual laborer in Jerusalem, according to Nawratzki, was two francs per day. On the optimistic assumption of three hundred working days per year, such a worker had an annual income of six hundred francs. Consequently, the annual revenue of a "typical" vintner in 1911 was at least one-fourth greater than the income of a Jewish manual laborer in Jerusalem. However, a "typical" citrus grower earned almost ten times more that year than the Jerusalem manual laborer and 7.5 times more than a vintner in the same colony.[25]

To earn this plenteous income, however, the citrus grower had to come up with the capital for purchase of the orchard land, prepare the parcel for planting, and cover expenses until the orchard began to produce enough

TABLE 4.1
Net Return and Profitability in Orchard Crops, 1911
(per Turkish dunam)

	Wine Grapes	*Oranges*	*Almonds/Apricots*
Land	45 fr[(1)]	150 fr	40 fr
Irrigation system	No	Yes	No
Trees per dunam	250	60	50
First fruition	4 years	5 years	5–6 years
Full fruition	6 years	7–8 years	7–8 years
Investment until first harvest	125 fr	1,000 fr	140 fr
Investment until full harvest	175 fr	1,200 fr	160 fr
Gross return per producing orchard	40 fr	220 fr	36 fr
Expenses per producing orchard	20 fr	100 fr	15 fr
Net return	20 fr	120 fr	21 fr
Return on equity (ROE)	11%	10%	13%

(1) Francs.

Source: Curt Nawratzki, *Die Juedicshe Kolonisation Palaestinas* (Muenchen: Reinhardt, 1911), 354.

fruit to generate revenue. This considerable sum created a barrier to the expansion of citriculture during this time. Growers such as Meir Apfelbaum considered that a blessing; after all, it would impose "natural limits" on the growth of orange supply. Even if prices were to fall by "up to 25 percent," Apfelbaum claimed in 1913, "citriculture would still reward its practitioners handsomely."

Summing up, most of the new Jewish colonies in the Judea area were clustered in an area fifteen to twenty-two kilometers outside of Jaffa. The distance from Jaffa made haulage to and from the city more expensive, and the distance from the Mediterranean made groundwater deeper down than along the coast. The large expected net revenue per dunam of orchard must have been a sufficient incentive for the investment despite the expected increase in haulage expenses. However, the technological bottleneck of pumping from the low water table, as stated, was the main impediment to expansion of citriculture dunamage. Due to the bottleneck, those who wished to invest in an orchard in the colonies of Judea had to spend much of the money on acquiring and installing a modern water-pumping system. Aaronsohn and Soskin estimated the investment in the pumping system at one-third of the total first-year investment; in 1911, Nawratzki estimated the cost of installing the pumping system at 57 percent of the first-year investment (neither estimate includes the investment in land).[26]

The high investment barrier did more than limit the number of investors in citriculture; it also induced efforts to improve profitability. The goal could be accomplished by widening the spread between net revenue and total expenses, that is, by cutting expenses or increasing output per dunam or—best of all—a combination of both. The main way to cut expenses was by increasing orchard size and spreading the investment over the largest possible land area. This is because almost the entire outlay for the pumping system was a fixed expense. Therefore, one solution was to maximize the size of the orchard that the expensive system could irrigate. Indeed, as we have seen, orchards equipped with modern pumping systems were at least twice as large as the others. Several growers in Petah Tikva, as noted above, formed investment groups for the preparation and planting of large orchards, such as the Bahariyya. This reflects the importance of scale economies, which growers at the time repeatedly mentioned as a possible way to improve the profitability of their orchards by cutting expenses and enhancing efficiency.[27]

THE SITUATION ON THE EVE OF WORLD WAR I

The well and irrigation system: The Jewish colonies' adoption of the internal combustion engine and mechanical devices for pumping was discussed above. Here we make brief note only of such changes as occurred up to 1914. The drilling technique was improved, but growers continued to sink their wells to groundwater because it was there that the pump attained its maximum capacity. Most pumps at the time were suction pumps of the piston type, which had to be as close to the water line as possible because their lifting capacity was limited to 8.5 meters.

Now, too, it was no longer necessary to sink the well at the highest part of the orchard, since the pump delivered the water to an elevated pond. Furthermore, cement replaced plaster as a sealant for irrigation ditches in the orchard. However, not in all orchards were the irrigation ditches lined with stones or sealed with cement. Capital-starved growers settled for sealing main ditches and left secondary ones unsealed.[28]

Land preparation took place at this time by a combination of manual deep hoeing (bahar) and mechanized cultivation methods, chiefly disking and chiseling by means of implements hitched to horses or mules. Bahar remained in use, it seems, due to the relatively low cost of labor as against deep ploughing by means of manège or a locomobile, and due to the growers' belief that bahar was the best long-term way to eliminate couch grass and other weeds. By the same token, continued loosening of soil and weeding seemed no more costly when done by shallow ploughing and mechanical disking than by manual hoeing. It is true that the grower and economist Meir Apfelbaum, in a detailed article published in 1912–1913, showed that light ploughing for ongoing cultivation of the orchard was preferable to the same action by hand.

Apfelbaum spoke of "an average orchard of six hectares" (60 dunams), a large area in terms of orchards of the time. This is unequivocal evidence of the scale economies that growers obtained by planting larger orchards than the local norm; it also provides empirical support for researchers' models of the dissemination of technological innovations, in which the adoption of the innovation increases commensurate with the size of the farm unit.[29]

Spacing: By this time, Aaronsohn and Soskin had already recommended and described a wider spacing of trees than the local norm. In 1913, Apfelbaum noted this as an accepted practice in Jewish citriculture. Wider spacing made it possible to use mechanical implements such as cultivators and disks between rows, thereby lowering cultivation costs, especially in large orchards. Generous spacing also had advantages in yield. Branches of trees did not become intertwined as in the traditional orchard and had greater access to sunlight and air. Thus, modern orchards outperformed traditional orchards in yield per tree. Furthermore, wider spacing proved to be profitable in the long term, because the density of planting in traditional orchards caused the trees to weaken and lose parts of their crowns after several years. A widely spaced orchard left enough space for each tree to be long-lived and, in particular, to continue producing even after its crown grew and spread.

The drawbacks of wide spacing were also economic: (1) the wider the spacing, the longer it took for the investment to mature. In spacing of 3.5 meters, maximum revenue was obtained ten years after planning; in spacing of six meters, it took twice as long, about twenty years, to maximize revenue. Obviously, too, when the increase in spacing was overdone, the negative effect of the decrease in number of trees nullified the positive effect of the increased yield per tree. (2) Another disadvantage of wide spacing pertained to the conditions of Palestine at the time: cheap labor as against scarcity of land and high cost of mechanization and its use. Here, too, scale economies played an important role in the extent of widening that a grower could choose. These circumstances, in my opinion, disprove the argument advanced by Shaul Katz, a researcher of technological development in Jewish agriculture in Palestine, that agricultural know-how is a composite phenomenon and that, consequently, its adoption in Jewish agriculture at the time was undertaken in isolated steps.[30] A technological innovation of seemingly miniscule value—wider spacing—led to additional changes: mechanization of cultivation and chemical and organizational improvements in technology. Furthermore, the more intensive the cultivation and the greater its capital component, the more complex the structure of the know-how became. The growers needed the advice of an expert such as Apfelbaum to keep the components of citriculture knowledge in balance.

The spacing that Apfelbaum recommended—"the best [spacing] for sunshine and air and the one that allows the greatest profit per hectare"—was four meters in light soil and five meters in heavy soil. Several years later, in

1919, another source attested that the conventional spacing in Jewish orchards was smaller than that recommended by Apfelbaum, at 3.5 meters (as against 1.5–2 meters in the traditional orchard), and that most cultivation was manual. Thus, the majority of growers, who did not participate in ownership of a spacious orchard such as the 650-dunam Bahariyya (in which Apfelbaum was both a co-owner and the labor foreman), preferred dense spacing and the earliest possible harvest of consequence. If so, the questions of profitability, level of investment, and duration of return on equity—crucial factors in the decision to adopt a technology, as the research literature corroborates—figured importantly in the growers' decision to adopt only partly the technological innovation of wider spacing, in view of the high cost of land and mechanization as against the abundant supply of labor.

As they widened their spacing relative to the practice in traditional citriculture, the growers also introduced fertilization as a systematic part of the cultivation process. In addition to organic fertilizer, which was bought mainly from Arab villages, and in addition to chemical fertilizer, "green fertilizer" (mainly lupine and clover) was planted to add nitrogen to the soil. Wider spacing made fertilization easier to perform and facilitated the introduction of mechanization such as disking in the fertilization processes.[31]

Propagation: In the traditional propagation method, as explained above, the process took about four years to reach the grafting stage. The method was comprised of the planting of shoots in nursery soil for two years in order to develop sturdy saplings, transplanting the saplings in the orchard, and waiting another two years until they developed a large enough root system to serve as rootstock for the grafting of a Shamouti scion. The Jewish growers, using the agronomic know-how that Baron de Rothschild's experts had imported, adopted a radically different propagation method that halved the time from the beginning of propagation to the grafting phase. First, instead of shoots, seminal propagation was introduced. Shoots have the advantage of maintaining the characteristics of the parent plant, but seminal propagation is less expensive and produces a much wider selection of seedlings. Additionally, saplings grown from seeds spent only one year in the nursery. Afterward they were taken to the orchard, where the grafting was done a year later. Thus, the period of time from germination to grafting was two years. The growers' purpose in shortening the grafting time relative to traditional citriculture was to obtain the first significant harvest sooner. The sources indicate that this had no adverse effect on the productivity or the sturdiness of the tree.

Unlike the propagation method, the type of rootstock most commonly used in traditional citriculture, the sweet lemon, was not replaced to any perceptible extent. The main reason for this was the steady, abundant, and early productivity of the Shamouti orange when it was grafted onto the sweet lemon. However, Apfelbaum noted that some growers preferred the sour orange (known locally as *hush-hash*) as their rootstock, finding it better able to withstand trunk and root-collar rot. Apfelbaum's remarks refer to one of the best known and most common citrus diseases of the time, caused by overirrigation and, especially, basin irrigation. The use of sour orange rootstock was a recommended method of avoiding the problem. However, as Apfelbaum noted, the Shamouti's yield is later and less regular when it is grafted onto sour orange rootstock than onto the sweet lemon. Another issue was the height of the graft. To prevent rot, some authorities, such as Peretz Pascal (a well-known grower from Petah Tikva, who acquired his expertise in horticulture through his employment in Baron de Rothschild's administration), recommended not only the use of sour orange rootstock but also a high graft. This, however, resulted in a later yield. Accordingly, despite the risk of losing some trees to rot disease, growers preferred to use low grafts and make the trees productive as early as possible.

In traditional citriculture, income covered expenses only from the beginning of the seventh or eighth year, as we have seen, or in the seventh year, as Apfelbaum reckoned. In contrast, other contemporary testimonies indicate that Jewish citriculture reached the break-even point by the fifth year. The main difference between Apfelbaum's findings and the others probably has to do with spacing. Apfelbaum worked out a spacing of four meters but, as we have seen, Jewish growers tended to plant at intervals of 3.5 meters. This is why they attained a greater harvest from the fifth year on and began to recoup their investments earlier than Apfelbaum's reckoning shows.[32]

ORCHARD INVESTMENT AND PROFITABILITY

Table 4.2 shows the cost of investment per dunam of Jewish-owned orchard, up to and including the fifth year, on the eve of World War I. The costs of labor, capital, and land in this reckoning serve as a point of departure for further discussion of technological improvements that Jewish citriculture introduced during the Mandate era. Afterward, we calculate the profitability of a producing orchard during that time. This calculation, too, will serve as a point of departure for an examination of citriculture profitability during the Mandate era.

The table shows that if we add the cost of installing the well and allied systems to the capital component, capital was the most important input in orchard investment at that time, at 38.1 percent of total investment. Apfelbaum (on whose figures our calculations are based) does not itemize the cost of labor invested in sinking the well, instead including it in the aggregate that he calls "Well, Pool, Pump, and Motor." However, capital goods were needed in installing the well, in order to build the support system of the walls. As the work became more complex, even at this time, more use was made of capital goods to sink the well and install the pond. Furthermore, the investment in

TABLE 4.2
Investment per One Dunam of Jewish-Owned Orange Grove, 1913 (five years)
(Palestine pounds per dunam)

Component	*1913*	*In 1936 prices*	*In percent*
Land	8.00	7.00	14.0
Well and irrigation system	12.40	10.85	21.7
Labor	12.45	10.89	21.8
Other capital	9.40	8.22	16.4
Interest[(1)]	14.89	13.03	26.1
Total investment capital	57.14	49.99	100.0

(1) "Interest" is composed of interest on current expenses, at 4 percent on annual average, and interest on the investment over the entire five-year period, at 8 percent (compound).

The conversion to Palestine pounds was performed at an exchange rate of 25 francs = £1 (P£1 after 1927). The conversion to 1936 prices was performed in two phases: conversion to 1921 prices according to Hagit Lavsky, *The Foundations of Zionist Financial Policy: The Zionist Commission 1918–1921* (Jerusalem: Yad Izhak Ben-Zvi, 1981) (Hebrew), 60, and conversion into 1936 prices by calculating the calendar index back to 1921, in accordance with the figures in Nachum T. Gross and Jacob Metzer, "Public Finance in the Jewish Economy in Interwar Palestine," *Research in Economic History* 3 (1978), Table 2, Column 2.

Source: Meir Apfelbaum, "Citrus Growing in Palestine," *HaHaklai* 1 (1912/13): 116–18 (Hebrew).

installing the pumping and water-delivery system surpassed the cost of land by 50 percent or more, and it, too, had to be carried out in the first year. Interest also accounted for a large share of the total investment. Unsurprisingly, then, many sources from the time urge investors to use as much of their own capital as possible and not to rely on loans. Indeed, as we have seen, typical investors in the initial phase of Jewish citriculture were very well-heeled.

This brings us to the question of citriculture profitability at that time. In regard to investment, the most detailed computation is that of Apfelbaum in 1913. However, unlike his detailed and accurate computation of investment, Apfelbaum's reckoning of orchard profitability does not take all factors into account. The main lacuna is the cost of current financing expenditure and capital recovery. Apfelbaum does, however, present abundant and contradictory data on the yield of a producing orchard and the FOB (free on board) price per case. For example, in his presentation of orchard investment data he claims that a harvest of 120–150 cases per dunam was attained twelve years after planting. Sometime later, however, when he published data on the prices obtained for oranges at Jaffa Port in successive editions of *Ha-haklai*, he reported the average yield at only one hundred cases per dunam. He did the same thing in regard to the FOB price. Initially he reported the price at five to ten francs per case, but immediately he circumscribed this finding and reported "the average over past few years" at 5.3 to six francs per case, and his calculations cited the lower figure.

There seem to have been several reasons for this. First, Apfelbaum may have been influenced by the relatively low prices that growers received in the 1912/13 season. Second, the industry was in the midst of growth and had a mix of young and mature orchards. In the year that Apfelbaum published his survey, Jewish-owned orchards such as those of the ICA in Petah Tikva and the first orchards that had been planted in that town, were twenty years old and had been delivering full yields for more than eight years. Concurrently, fruit from younger orchards was being delivered to the market at Jaffa Port, and these orchards had lower yields than the mature ones and produced lower quality fruit that commanded lower prices. It is very likely, however, that Apfelbaum deliberately refrained from providing full information on the profitability of the industry. Partial disclosure of profitability data is a logical thing for a private investor to do; it was also reflected in the minutes of Pardess board meetings at the time and recurred during the Mandate era as well.[33]

Table 4.3 presents detailed profitability calculations at three levels. Level A is a minimum reckoning (the one used by Apfelbaum), B is an average, and C is a maximum at the production and price levels that Apfelbaum cited. Interest on current expenses and capital recovery were added to the reckonings at all three levels.

Apfelbaum's statement that "citriculture would still reward its practitioners handsomely" even if prices fell by 25 percent suggest that the lowest price that he cited in his reckoning, 5.35 francs per case FOB, cannot be the actual price that Jewish growers received for their fruit. A 25 percent decline from that level would put the growers in the red. Even if we, like Apfelbaum, exclude the cost of interest and capital recovery from our reckonings, the profit after a 25 percent decrease in prices would be only 1.08 P£ per dunam or a 2 percent ROE. Thus, the profitability of a fully productive orchard at the time was probably much closer to price B or price C. These figures are reinforced by Smilansky, who estimated the profit on capital invested in a producing orchard at 30 percent.[34]

Thus, the Jewish growers introduced several important technological innovations in Palestine citriculture in 1900–1914. Two main factors prompted the introduction of these innovations: rising demand for Shamouti oranges in foreign markets and expectations of a continued uptrend in profitability. The agro-technical nature of the changes was based on the general agronomic knowledge of Baron de Rothschild's experts and not on referral to

TABLE 4.3
Profit and Loss per Dunam of Productive Jewish Orange Grove, 1913
(Palestine pounds per dunam)

Component	A	B	C
Cultivation expenses	6.00	6.00	6.00
Packing materials	6.24	6.50	7.80
Picking and packing labor	2.16	2.25	2.70
Haulage from Petah Tikva to Jaffa	0.96	1.00	1.20
Miscellaneous (loading, stevedores)	2.40	2.50	3.00
4% interest on the foregoing	0.71	0.73	0.83
Total expenses	18.47	18.98	21.53
Capital recovery at 8% for 30 years	5.07	5.07	5.07
Total costs	23.54	24.05	26.60
Revenue FOB	25.68	37.50	60.00
Net profit per dunam	2.14	13.45	33.40
Pct. profit on equity invested (over 8%)	3.7%	23.6%	58.4%

A = 5.35 francs per case FOB @ 120 cases per dunam (Apfelbaum's reckoning).
B = 7.50 francs per case FOB @ 125 cases per dunam. 125 cases per dunam is a reasonable average of the two levels (150 and 100 cases) that Apfelbaum cited.
C = 10.0 francs per case FOB at 150 cases per dunam—the maximum figures that Apfelbaum presented. Apfelbaum claims that some orchards actually generated this output.

Notes:

1. The reckoning is for a fully productive orchard. At Price A, the orchard begins to break even at the end of the seventh year after planting; at prices B and C, this occurs at the end of the sixth year. According to Apfelbaum, the maximum yield is attained in years 10–12 after planting. This is corroborated by additional sources, e.g., F. Sheinfeld, "Fundamentals of Orchard Evaluation," *Hadar* 10:1 (1937): 10–11. However, Apfelbaum, like all citriculture experts during the Mandate era, speaks only of the profitability of a fully productive orchard. It took four to six years to move from the break-even point to full yield. The duration of the transition depended mainly on the quality of cultivation, an inconsistent matter at the time. Moreover, as we see below, the growers always tried to shorten the transition period. Since our study does not intend to detail the cost of investment in each orchard, the table does not include reckonings for the transition period.

Source: see Table 4.2.

a specific model. The remarks of Apfelbaum and Smilansky reflect this clearly, and other researchers, especially Shaul Katz, have shown it to be the case. The increase in spacing, the use of fertilizers, the measurement of quantities of water for irrigation, and the war on pests and diseases (which was more potential than actual at that time) were the results of application of general horticultural knowledge in citriculture and not of special expertise in this industry, since such expertise had not yet developed. The introduction of these innovations also stemmed from the relationship between the supply of mechanical and chemical resources in Palestine and the growers' demand for them. Chisels and disks, manège and locomobile motors, and organic and other fertilizers were introduced in Palestine by various agents (the Baron's administrators, the Germans, and local developers) irrespective of the citriculture industry, which was a fledgling enterprise at the time. The growers put these innovations to use and demanded them in rising degrees as their industry grew. Demand for citrus in export markets and pressure to solve the water pumping problem, as we have seen, abetted the introduction of internal combustion motors and mechanical pumps.

The socioeconomic nature of the Jewish citriculture that evolved during this time, too—a capitalist industry with colonial economic attributes—also affected the gist of the technological changes that the industry adopted. Jewish citriculture developed in settlements (*moshavot* or *koloniyot*—"colonies"—as their contemporaries called them) populated by people of European origin who had moved into an economically undeveloped area and established there an autonomous society that had a penchant for capital and technology intensivity. On the one hand, the industry that they developed entailed large capital investments that assured a handsome return; what is more, the industry had a rather high share of absentee owners. On the other hand, the growers used the cheap unskilled labor of local inhabitants. Both factors, capital and labor, strongly affected the extent of the adoption of technological innovations in citriculture.

World War I placed Palestine citriculture in the throes of a grave crisis, from which the industry recovered only in the mid-1920s. The cessation of exports; the tough military and economic regime that the Turks imposed on Palestine and, especially, its Jewish sector; the shortage of fuel for irrigation of orchards; and a severe plague of locusts in 1915 combined to nearly wipe out the industry. The Jewish growers adapted their kerosene motors to make them run on wood and coal. By so doing, and by taking loans from the Anglo-Palestine Company (subsequently the Anglo-Palestine Bank), the ICA, and other creditors they sustained their orchards until the hard times passed. The resulting severe indebtedness was one of the main factors that impeded the recovery of the industry after the war.[35]

FIVE

Technological Innovations during the Mandate Era

BASIC FACTORS AND IDEOLOGICAL CONSIDERATIONS

The California Model

BEFORE GREAT BRITAIN received its mandate for Palestine, the Jewish growers there had been introducing technological innovations—relative to the point of departure, local Arab citriculture—on the basis of general agronomic and technical know-how that came mainly, and most importantly, from the agricultural experts of Baron de Rothschild's administration and the ICA. Later on, during the Mandate period, the Jewish growers found a different source of inspiration: a fully developed and economically viable system known as the "California model." Although the growers were aware of other citriculture systems, such as those used in Spain, South Africa, Algeria, Italy, and Florida, it was the California model that they, like growers in many other countries, adopted.

Several basic features of the California model made this method more attractive than the alternatives to the entrepreneurial Jewish citrus growers of Palestine. First, California was the world's largest citrus producer at the time. Second, the climate of California closely resembled that of Palestine; in California, for instance, orchards were irrigated in the summer and made do with rainfall in the winter. The main attractiveness of the California model, however, was its unique synthesis of advanced technology, centralized organization, and the integration of private entrepreneurship and government activity. Due to these factors, California citriculture recorded steady growth and provided private entrepreneurs with regular profits.

Rather than delving into the details of California citriculture and the diffusion of the California model among the Jewish growers in Palestine,[1] the following discussion stresses two main aspects only.

1. The California model was regarded as the epitome of a Western national society that had created autonomous governing institutions, highly developed education and research systems, a modern infrastructure, and profitable agriculture based on advanced technology, private entrepreneurship, and economic liberalism.
2. The California model was widely known in Palestine; everyone in the industry—not just experts—referred to it routinely and necessarily. Jewish citriculturists knew of it by means of growers' instruction manuals, public lectures, the press, visiting foreign experts (whose travel expenses were covered by the growers or the government), native-born Jewish farmers who were sent abroad to study citriculture, and Palestine agronomists and citriculture leaders who visited California to acquire firsthand knowledge of the industry.

As our discussion continues, the California model will resurface time and again as a benchmark against which growers in Palestine gauged the technological and organizational development of their industry.

Technological change in Palestine Jewish citriculture was not only an economic question; ideological considerations mingled as well.

In their attitudes toward technological innovation in Jewish citriculture during this time, both experts and growers fell into three main "camps." Those in the first camp proposed to transform the technology of Jewish citriculture in accordance with the California model. One may term them "modernists" because their motives were largely modernistic, that is, they regarded the introduction of modern technology as a value in itself and technological achievement as the path to a better future. They also viewed modern technology as the best and most appropriate way to cut costs and increase yields. The second camp was made up of "national modernists" and, especially, those of the socialist (Labor Zionist) persuasion. They, too, were modernistic in their worldview but their modernism had a socialist and nationalist overlay. For them, the introduction of modern technology in Jewish citriculture would figure importantly in leading the Jewish economy of Palestine—in which a citriculture was a mainstay—to independence from the Arab economy and Arab manual labor, and in creating a more just society. Those in the third camp—mainly private growers, particularly in veteran colonies—were skeptical about the utility of uncontrolled introduction of technological innovations. Although they did not totally rule out the values of modernism and Zionism, in view of the socioeconomic realities of Palestine they opposed the unrestrained adoption of every technological innovation discovered.

The main dispute surrounding the technologization of Jewish citriculture concerned the introduction of labor-saving technology for manual labor. Apart from being the paramount symbol of "modernization," this issue had palpable socioeconomic implications for unskilled laborers and interrelated with the national issue of "Arab labor." Disputation over the introduction of other innovations, such the biological and the chemical, was mainly economic; it was uncontaminated by social and national elements that evoked discord and emotion.

JEWISH LABOR OR ARAB LABOR?

The research literature has spoken at length about the futile struggle against Arab labor in the history of the Jewish settlement enterprise in Palestine.[2] It is not our concern here to delve into this issue once again. Instead, we will note only the main economic aspects of the issue. One important facet was the difference in labor costs. Although the prime movers of Histadrut citriculture (such as Yehuda Horin, manager of Yakhin) attempted to prove that Jewish labor was not much more expensive than Arab labor, all other sources—growers, independent bankers, contemporaries whose writings were not meant for publication, and studies in our time—show clearly that Arab labor, both in routine cultivation and in harvesting and packing, was at least 40 percent cheaper (per day) than Jewish labor. Furthermore, apart from the wage differential, most growers repeatedly complained that Jewish

FIGURE 5.1. Jewish demonstrators outside a private-owned orange grove in Kfar Saba, 1934. Among them the famous Zionist poet, Shaul Tchernichovsky.

workers put in fewer hours (eight as against twelve by the Arab laborer) and were much less productive.

Anita Shapira's research on this topic (referred to above) shows that Jewish labor never accounted for less than one-third of total labor. Thus, most citrus growers employed quite a few Jewish laborers alongside Arabs. Their motives in doing so were complex. A veteran grower in Rehovot, such as Tovia Miller, was averse to employing Arabs in his orchards for national reasons. Growers in the Sharon region, who organized under the umbrella of the National Farmers' League and also fought the battle for Jewish labor within the Farmers' Federation, were even more emphatic about this. Security considerations and social and cultural affinity also prompted many growers to employ Jewish labor collectives in their orchards. It was generally assumed that Jewish immigrants would gravitate to the colonies, where they would receive assistance from the governing committee, obtain their initial source of livelihood in Palestine, and find a supportive host society and a familiar culture. Memoirs and testimonies from the time show clearly that the picture of dichotomy and hostility between private farmers and farm workers (most of whom belonged to the Labor Movement population), as portrayed in the propaganda of the time and, by percolation in some historiography, was only one aspect of the matter, and a non-representative one at that. Sympathy for the laborers ("workers") and the roles they played in the Zionist enterprise, as well as social relations and marriages between the offspring of farmers and laborers in the colonies, were only two elements in a historical reality that was much more harmonious than the one described in the propaganda and some of the historiography.[3] This is not the place for an extended discussion of this fascinating issue. By mentioning it, we merely note that the relatively high proportions of Jewish labor and, a fortiori, of mixed Jewish and Arab labor should not be surprising against the social, national, and cultural backdrop of life in the colonies.

It is hard to quantify the effect of the more expensive Jewish labor on the profitability of Jewish citriculture. The statistics that address this question are incomplete and poorly grounded. Clearly, however, it was less costly in Arab citriculture than in Jewish citriculture to invest in planting an orchard up to the first significant harvest and to cultivate productive orchards on a regular basis. According to Metzer and Kaplan, Arab citriculture performed these activities at 60 percent of the cost in Jewish citriculture.[4] However, data from the end of the period at issue show that the disparity diminished over time, especially in respect to the costs of maintaining a productive orchard, including harvesting and packing. For example, Moshe Smilansky wrote in his discussion of the crisis season of 1936/37 that "For the Arabs, the expenses per crate are one shilling less—the labor difference." According to his reckoning, total current annual expenses per crate of oranges in Arab citriculture were nearly ten shillings (nine shillings and eight pence). If we accept his calcula-

TABLE 5.1
Profit and Loss per Dunam of Productive Orchard, with Jewish Labor, Arab Labor, and Mixed Labor, 1932
(Palestine pounds per dunam)

Component	*Jewish labor*	*Arab labor*	*Mixed labor*[(1)]
Cultivation	6.00	3.60	4.56
Harvesting and packing	7.20	5.64	6.30
Materials and misc.	10.00	10.00	10.00
4% interest on current expenses	0.93	0.77	0.83
Capital recovery, 30 years, at 8%	9.11	8.53	8.76
Total costs	33.24	28.54	30.45
Turnover at orchard gate	43.20	43.20	43.20
Profit	9.97	14.66	12.74
Difference (%)	—	47	28

Source: Yakir Plessner, *The Political Economy of Israel: From Ideology to Stagnation* (Albany: State University of New York Press, 1994), 174–76.

(1) Forty percent Jewish labor. My computations of the cost of capital recovery differ slightly from Plessner's, evidently due to a difference in calculating investment capital.

tion, the difference in annual current labor cost between Hebrew and Jewish citriculture hardly surpassed 10 percent.[5]

However, as stated, the norm in Jewish citriculture was mixed labor. Table 5.1 illuminates two interesting facts: the small proportion of labor relative to capital goods and the price of capital, especially in investing in orchards but also in harvesting and packing, and the difference in profitability between an orchard cultivated solely by Jews and one cultivated in the normal way, by mixed labor. This difference was not 47 percent, as in the difference between the two eitheror alternatives, but only 30 percent. By appreciating this, we may better understand the growers' struggle against raising the wages of the Jewish workers. The large share of fixed expenses in costs, on the one hand, and the reality of mixed labor, on the other hand, left a small margin within which profitability could be maintained when turnover prices were constant, let alone when they were falling. Labor costs were quite important in maintaining this margin.

THE IMPLEMENTATION OF MODERN TECHNOLOGY, 1926–1939

This section explores several main aspects of the implementation of technological innovation in citriculture at two points in time: the late 1920s, when the "planting frenzy" began (1926–1931), and about ten years later, when citriculture tumbled into crisis and planting stopped (1936–1939). Naturally enough, the main issues in the late 1920s concerned the cultivation of young orchards and the emphasis shifted in the late 1930s from the young orchards (which had matured in the meantime) to mature orchards that produced full harvests.

A Tale of Two Paths: Old Citriculture and New Citriculture

In the late 1920s, the main question in land preparation was whether to continue using the traditional bahar method or go over to deep ploughing. When the "planting frenzy" began, much larger parcels had to be prepared than before, in view of the tremendous scale of the planting enterprise under way. In 1927 the American expert Knowles Ryerson, who visited Palestine to examine the domestic citrus industry as a member of the Zionist Organization delegation of Experts, noted that

> deep plowing is the universal practice in preparing land for citrus groves. According to Government figures, 113 tractor outfits of varying capacity are in operation for this purpose in Palestine at the present time. The depth varies with the size, the most thorough work being accomplished by steam or caterpillar tractors using cable plows. With this equipment a furrow up to 40 inches [over one meter] in depth is turned leaving the soil in excellent shape for root growth.[6]

Ryerson expressed doubt about the need for deep ploughing, arguing that good results could be attained at lesser depth. The agronomist Melekh Zagorodsky, manager of the Merhavia cooperative for some time and an employee of Zionist Organization Agricultural Settlement Department during the Mandate era, was also well acquainted with the citrus industry. In a special manual for growers that he wrote in conjunction with two industry leaders, Tovia Miller of Rehovot and Peretz Pascal of Petah Tikva, he presented a more complex picture:

> Today [1928], at a time of transition and a war of methods, we find different labor arrangements in various places. Many behave in the following way: they plough with a tractor or with a large number of animals; laborers follow the plough and gather into tins or baskets the couch grass [Bermuda grass] that the plough has managed to discover and uproot, and in the spring, after individual blades of couch grass come up, they are removed with hoes, digging as far down as the roots go.
>
> Generally speaking, it is a question of financial resources; the first method—tilling and destruction of couch grass by hand—is appropriate for small and medium orchards, and the second method—deep ploughing and cultivation to destroy the grass—is appropriate for large orchards and for expert and well-heeled owners.[7]

Did this situation change by the end of the period? In 1938, Y. D. Oppenheim, manager of the Citrus Plantation Department of the Farmers' Federation, wrote that Ryerson's doubts about the need for deep ploughing "have also infiltrated farmers' circles. However, [the farmers] were influenced mainly by the high cost of labor and began to settle for shallower ploughing to prepare the land." If in 1928, according to Zagorodsky, the proponents and opponents of bahar were still waging a "war of methods," ten years later Jewish citriculture had almost totally stopped preparing orchard land by hand. The reasons were the high cost of manual hoeing and a steep decrease (67 percent on average) in the cost of deep ploughing per dunam. The difference in ploughing depth certainly had a powerful effect on cost. At the beginning of the period, ploughing methods such as bahar went down to between eighty and one hundred centimeters; in the late 1930s the plough turned soil to a depth of fifty centimeters on average, if not less. Deep-ploughing costs were also cut by the mammoth increase in planting area, especially in the Sharon region.[8]

In the new plantation areas of Sharon and Samaria, where little citrus planting and Jewish settlement had taken place before the late 1920s, we find a much greater propensity than in the veteran colonies to adopt new technologies in land preparation and subsequent phases of planting and cultivation, such as wider spacing and the installation of the "California" irrigation system (matters that are discussed below).

What was the root of the difference between the new areas and the earlier places of agricultural settlement? As it seems, the tendency to adopt technological innovations in the Sharon and Samaria traced to special characteristics of the settlement method used in those regions. (1) The intended orchard area was largely unsettled and was therefore conducive to careful planning. (2) In many cases—the orchards of Tel Mond, Netanya, and the Yakhin areas, to name only a few—planting area was extensive. Even if the orchard area was eventually subdivided into small parcels (as happened in most cases), scale economies came into play in land preparation, planting, installation of irrigation infrastructure, and other works, bringing down costs relative to those in smaller land areas. (3) Most developers had come to Palestine from the West and, therefore, were aware of Western technological innovations. (4) The issue of Arab labor and the wish to minimize it in the Jewish settlement enterprise lurked in the background.

As stated, settlement in these regions, especially the Sharon, was largely organized and planned in advance. In some cases, large developers acquired a large parcel and readied it for cultivation and a large potential population. An example is Tel Mond, where Lord Melchett (Alfred Mond) bought land and prepared it for the cultivation of citrus orchards that would be sold off. In other instances, groups came together in advance of settlement—some did this before they reached the country—and bought and prepared the land in a centralized manner. To carry out these actions, and even to cultivate the orchard until bearing age, the new settlers contracted with local firms that had been established for this purpose. Notable among these companies were Hanotea, Ltd., founded by the offspring of farmers from the old colonies; a company headed by Colonel Frederick Hermann Kisch (former director of the Jewish Agency Political Department and afterward a private businessman); a company headed by Moshe Smilansky of Rehovot; and the Histadrut-owned Yakhin enterprise.

Citrus areas were prepared on the assumption—in which many experts concurred—that by applying "rational" methods in development, irrigation, and cultivation itself, the grower and his family might cultivate a twenty-dunam mature parcel with no assistance save several weeks at harvest time. Such an orchard might provide its owner with a decent income. Plans for the Keren Hayesod farms and those in the Hefer Valley, in contrast, envisioned much smaller orchards, at only ten dunams. In these cases, the preferred model was mixed farming, which would avert the riskiness of single-crop production and meet most of the farmers' household needs.[9]

An important example of the preparation of orchard groves for and by private developers was Hanotea, Ltd., established in 1929 when settlement in Netanya began. The founders explained in so many words that they had no intention of buying small parcels of land in various parts of the country, where main transport routes and Jewish centers would be inaccessible. Their express intent was to create large settlement blocs that could be managed and cultivated by means of "modern settlement methods . . . and scientific methods." Hanotea established a technical department with a laboratory for soil testing, employed agronomists, and based itself on the vast experience of its members, who, after all, were the offspring of farmers and citrus growers. Thus, the plantings in Netanya made use of modern technologies that exploited scale economies: deep ploughing by caterpillar tractor instead of bahar, "California irrigation" (explained below), and deep wells with electric pumps. The company also promised to establish large mechanized packing facilities. Hanotea then established large citrus plantings in additional areas—Even Yehuda, Rishon Lezion, and Nana, not far from Rehovot—using similar methods.[10]

A pamphlet put out by Yakhin in 1930, marking the company's fourth anniversary, gives a typical description of the intensive mechanization that the company used in preparing and cultivating orchard land. Yakhin, which championed technological innovation as an ideology, strove to show that its technological achievements were a proof of the economic efficiency of the "rational" cultivation method: (1) Not only had bahar been replaced by deep ploughing, but additional ploughing was done in the winter, when the land was softer; this, the company said, reduced costs. (2) The land was leveled before planting on the basis of a topographical map, resulting in an irrigation pattern that created a "convenient downward flow of water." The work was done by leveling equipment that the company itself had improved; thus, it was performed "quickly and well" and at a saving "of up to 30 percent of labor expenses."(3) Instead of digging eighty-centimeter pits for planting, the company dug running furrows at a depth of seventy centimeters up and down the orchard area. Thus, the land was prepared for planting without manual cultivation and the trees were planted "without the excavation of pits," thereby apparently sparing the considerable manual labor that went into the preparation of planting holes. (4) A "California"-style grid of irrigation furrows was laid out, not only saving money, the company claimed, but also allowing "animals or tractors to cultivate the whole area." (5) In the spring ploughing, after the rains and in advance of the irrigation season, the plough was replaced by a disk. This, Yakhin said, did a better job of softening the soil and inserting lupine ("green fertilizer") between the rows of trees and was cheaper than the use of a plough. (6) The *turiyya* (a short-handled hoe) was replaced with a long-handled implement that allowed its wielder "to work standing up and be twice as productive." The result was an increase in productivity and a contingent of satisfied workers, who worked as contractors of Yakhin.[11]

SPACING (PLANTING DISTANCES)

"No single question connected with young orchards is causing more comment than as to what is the proper planting distance for the Jaffa Orange," Ryerson stated in 1928, reflecting the Jewish growers' dilemma in regard to the integration of advanced technology in citriculture. Since Ryerson considered mechanization "part of the march of civilization," it comes as no surprise that his report on Palestine citriculture included an unequivocal recommendation to space the trees farther apart. The American expert listed several factors that, in his opinion, warranted this reform. (1) Severe density limits the productive area of the tree to the upper branches, since the lower ones are shaded and cannot produce fruit. The advantage of more trees per dunam did not compensate for the loss of producing branches. (2) A wage increase is inevitable, in view of rising standards of living in Palestine and an expected increase in demand for labor. Therefore, the growers should use wider spacing as a preemptive measure to facilitate mechanized cultivation. (3) Wide spacing facilitates suitable pest control. For these reasons, Ryerson considered the replacement of manual cultivation with mechanized cultivation inevitable. "Where 6 to 8 people are needed per acre [four dunams] now in Palestine, one man with seasonal help at picking time, fumigation, etc., can care for 10 to 20 acres, using modern methods."[12]

A salient reflection of the potential utility of technology-intensive citriculture was provided by a South African citrus expert, Professor H. Clark Powell, who was invited to Palestine by the Mandatory Department of Agriculture and Forests and published his findings in January 1928. In his report, issued as a special publication of the Agriculture Department and quoted lavishly in publications of contemporary citrus growers and experts, Clark Powell recommended a more generous spacing of five or six meters. This, he said, would improve the yield of each tree considerably and would slash production expenses. The improved yield, Clark Powell claimed, would more than compensate for the loss occasioned by fewer trees per acre. An orchard planted at 3.5 x 3.5 meters would have about one hundred trees per dunam, he said, and if the average production at this spacing were about one case per tree, the orchard would produce one hundred crates per dunam. If a spacing of 5.5 x 5.5 were adopted, there would be thirty-one trees per dunam. At such spacing, Clark Powell explained, the yield would rise to four cases per tree, that is, 125 crates per dunam, 25 percent more than the conventional spacing in Palestine would allow. He admitted that the fruit would ripen later at the wider spacing than in dense planting, but this drawback would be surmounted over a period of ten to twelve years. Notably, Clark Powell did not discuss the requisite investment in mechanization and beasts of burden; he may not have been fully aware of the low cost of human labor (foremost Arab) in Palestine. Despite these drawbacks in his calculus, many new Jewish growers seems to have accepted his point of view; in the planting period of 1926–1931 there was a general tendency to use a spacing of 4.5 x 4.5 if not more (41–52 trees per dunam).[13]

By 1932, however, when the four-year "big planting period" began, it was clear that growers had more to lose than to gain by adopting wide spacing. The model described above was inspired by California citriculture, in which vast areas of land were available and production was calculated on a per-tree, not a per-acre, basis. In Palestine, production per dunam was the main criterion and an early return on equity was preferred (for reasons including high interest rates). Thus, the situation changed during the "big planting period": the spacing was tightened to 4 x 4 meters (52–66 trees per dunam). Yakhin, too, which had originally recommended the full adoption of the modern model, stopped planting at spacing greater than 4 x 4.5 in the mid-1930s and admitted that wider spacing would not compensate for the decrease in yield per dunam.[14] Evidently, then, the computations of Ryerson and Clark Powell, which most experts accepted at first, were not fully appropriate for Jewish citriculture. The attempt to transplant the California model to Palestine and its conditions did not turn out well, mainly due to differences in conditions: scarcity of land, availability of cheap unskilled labor, high interest rates, and the wish to obtain a return on equity as quickly as possible. Nevertheless, the average spacing in the late 1930s, 4 x 4 meters, was wider than the late 1920s norm. Apparently the wider spacing also did much to improve production in Jewish orchards, which outperformed Arab orchards according to most estimations, and to reduce cultivation expenses.

FIGURE 5.2. Decoville in large Jewish grove, late 1930s.

ROUTINE CULTIVATION

Nearly all routine cultivation of citrus orchards—young and producing—took place between March and October, eight months per year. The main cultivation activities were winter hoeing for weeding purposes, irrigation, manuring, and treatment of the tree and its crown (pruning and whitewashing). Increased spacing in Jewish citriculture facilitated the mechanization of cultivation in young orchards. Thus, in the new planting areas—and in new plantings in older localities—mechanization was used until the orchard began to produce (five to six years) and until the canopy became too dense. Where wide spacing (5 x 5 meters) was invoked, mechanized cultivation could continue afterward. In most orchards, however, and especially in the older ones, most if not all cultivation was manual.

At the beginning of the Mandate era, a peak period in the profitability of citriculture, not only was nearly all cultivation of mature orchards manual but the growers even tended "to increase expenses and, thereby, revenues."[15] From the early 1930s on, however, as the profitability of citriculture declined steadily, the Jewish growers began to seek ways to reduce their cultivation costs.

One idea was to mechanize cultivation in dense orchards. In 1936, the agronomist B. Gorstein, manager of the Yakhin plantations department, proposed that decovilles be used in such orchards. The decoville was a low, shallow gondola that was propelled along a narrow-gauge track by people or animals. Its main use was in hauling picked fruit to the packing shed on the orchard grounds, but it was also useful in spreading manure. Five years earlier, in 1931, Moshe Smilansky recommended the introduction of the decoville in Jewish citriculture as a technological innovation that might justify its investment. Decovilles were not widely used, however, mainly because only large and widely spaced orchards had room for the tracks.[16]

Still, the Jewish growers definitely tended to cut back on manual cultivation at this time. They also reduced the quantities of organic fertilizer that they spread around their trees and invested less care in the crown, after learning that the tree grew to the same shape and delivered a similar yield without intensive treatment.

The total cost of cultivation in productive orchards declined somewhat, most significantly decreases in taxation (75 percent) and labor (32 percent). One reason for the steep decline in taxation was a reduction in rural property tax on farmland. In late 1937, after growers applied pressure due to the losses they incurred in the 1936/37 season, the tax rate was lowered by 40 percent, from 825 Palestine mils per dunam to five hundred mils. The decrease in labor cost shows that the downscaling of cultivation did result in considerable savings for the individual grower. It is also noteworthy that the share of labor in the total cultivation cost (labor plus some fixed expenses) slipped from 48 percent in 1927 to only 37 percent in 1938.[17]

Pumping and Irrigation Systems

A second focal point of technological innovation in Palestine Jewish citriculture was the water pumping and delivery system. As shown above, the main technological innovation in the Ottoman period was mechanized pumping that allowed the use of deeper groundwater. During the Mandate period, both major components of the system, pumping and irrigation, became more complex.

We use the expression *pumping system* to denote the system that delivers water from a natural source to the irrigation system in the orchard; the irrigation system carries this water to the trees. The main components of the pumping system were the pump and its power source; the irrigation system was usually composed of a small impoundment pool and a network of tubes and/or ditches in the orchard. The stronger the pump and its flow, the greater the grower's ability to introduce additional technological innovations (an underground tubing system, for example). However, an efficient and powerful pumping system did not make it *necessary* to introduce technological innovations in the irrigation system. Various factors affected the structure and technological nature of these two principal elements of the orchard water system.

Pumping Systems in the Late 1920s

Palestine/Israel has two main sources of water for irrigation: surface water and groundwater. During the period at issue, the main surface water sources were the Jordan River and the Yarkon River. However, they irrigated only a small proportion of orchard area countrywide. Only three thousand dunams of orchard out of 42,000 dunams in 1926 (7 percent) were irrigated from the Yarkon. (This accounted for most orchard area irrigated with surface water, since the citrus area in the Jordan Valley was very small.) The remaining 39,000 dunams (93 percent) were irrigated from wells. A decade later, at the end of the "planting frenzy," the industry was as dependent on groundwater as before. In 1937, Shmuel Yedidya, manager of the plantations department at Mikve Yisrael Agriculture School, estimated that more than 90 percent of orchard area countrywide (more than 270,000 dunams out of 300,000) was irrigated from wells.[18]

Four main factors determined the technical structure of the pumping system in Jewish citriculture. Three of them were discussed above: (1) the depth of the water table, (2) the cost of installing and operating the system, and (3) the availability of the technology used in the system. Such technology existed both in Europe and in California and was known to all Jewish and many Arab growers. Thus, its introduction was a matter of adoption; there was no need to invent it. The fourth factor was the availability of sufficient nonsaline groundwater on the coastal plain, the main citrus area in Mandatory Palestine. Although large-scale pumping caused the water table to sink steadily during the Mandate era, the developers acted on the assumption that water would always be available.

In 1927–1929, when the "planting frenzy" began, the average water table on the coastal plain was between twelve and thirty meters down, although water was also found at depths of between five and sixty-five meters. Thus, the traditional well and the old pumping facility, the waterwheel, could still be found in Palestine, especially in young Arab orchards. However, as stated, the traditional well could not meet the rising demand for water, especially when planted area increased by hundreds of percent in the years following the British conquest. In Jewish citriculture, the well was replaced by modern pumping systems that were established in individual orchards by individual growers, except for a few centralized water companies that were established in 1926–1929. Moreover, the availability of groundwater at a depth that still allowed growers to sink wells using cheap manual labor (mainly Arab) helped growers to remain autonomous in regard to water supply.[19]

Two main types of pumps were used at the time. The first was a suction-force pump, installed at the bottom of the well as close to the water table as possible. These were horizontal-piston or centrifugal pumps that operated in two phases: suction of groundwater to the level of the pump and forcing the water from the pump level to the orchard area. To install these pumps, it was necessary to sink a well that would reach the water table. The second type of pump used the principle of force only. A narrow hole was drilled from the surface to the water table, a tube with a pump at its head was threaded down the hole, and the pump forced water directly to the orchard. Such mechanisms were known as drill pumps.

The main criterion in the decision about the type of pump to install—a decision that determined the technological structure of the pumping system—was the cost of installing and operating the pump and the nature of its power source. The sources available to use show that it was not the flow of the water source that determined the size of the area intended for cultivation but the other way around. The idea that groundwater was an unlimited factor input was widely accepted and experience had even shown it to be justified. The availability and quantity of water were treated as constants. The only question was how best to get to the water in terms of technology. Therefore, when growers encountered a water source that delivered a weak flow, the obvious recommendation, in most cases, was to find an additional one.[20]

In 1927–1930, the most common type of pump was the suction-force type. The two components of the action of the pump, suction and force, imposed various limitations on installation. The suction principle limited the pump's lifting ability to six to eight meters under the water table. Thus, to install a suction pump one had to sink a well almost to groundwater in order not to waste crucial suction height. Therefore, a concrete floor was built at the bottom of the well and the pump was laid atop it. This placed the pump as close to the water table as possible and created a stable surface for its operation. Obviously, insofar as it became necessary to deepen the well in order to reach the water table—due to objective conditions related to the structure of the aquifer in a given area or after overpumping lowered the water table—the cost of installation escalated.

Most pumps for orchards obtained their power from one of two sources: diesel motors and electricity. Most submerged pumps were electric; those installed over the water table used either electricity (horizontal centrifugal pumps) or internal combustion motors (piston pumps).

How widely used were these types of pumps in Jewish citriculture in the late 1920s? Of 380 pumping systems in Jewish colonies in 1929, 320 (83 percent) were powered by internal combustion motors and sixty-five (17 percent) used electric motors. A year later, in 1930, there were 430 wells in the Jewish sector and 26 percent of them (112) were powered by electricity.[21]

Thus, electricity was a little-used source of pumping power at the time, although its share was rising.

One reason for the clear preference of liquid-fuel pumps was the incremental development of the country's electrical system, which in 1930 was in its infancy. The electric system used by the Jewish sector developed in three main centers: Tel Aviv, Naharayim, and Haifa. The first power plant in Tel Aviv was established in 1923 and had little capacity. Its output increased over the years, and a second power plant in Tel Aviv, a steam-turbine facility, was dedicated in 1938. The first plant in Haifa was established in 1925; it, too, expanded over time. About ten years later, a second and much more powerful power plant was built in Haifa, powered by three steam turbines. The plant in Naharayim, inaugurated in 1932, supplied about half of total electricity consumption countrywide by the end of the period at issue here, insofar as research addresses itself to the topic. Petah Tikva was connected to the electrical system in 1926, shortly after Tel Aviv. Since data about the rest of the electricity grid are fragmentary, it is difficult to draw the contours of the system with precision. However, available sources about the emplacement of the electricity grid for Jewish citriculture indicate that from 1930 the Sharon region was connected at a quickening pace and that electricity was extended to the other citriculture areas as well.[22]

If so, how was the gradual development of the country's electricity system reflected in the use of electric pumps in Jewish citriculture? In the settlements of the Judea area—the bastion of Jewish citriculture, where 85 percent of the total Jewish citrus area was cultivated at the time—only 22 percent of wells were powered by electric pumps in 1928. In Petah Tikva, where 43 percent of the Jewish orchard floor was situated in 1927, only a small fraction of pumps were electric. In the Samaria region (Hadera, Karkur, Pardes Hannah), however, electricity powered 59 percent of the wells (forty-five of seventy-six) in the late 1920s. In the Sharon region (Netanya and Tel Mond), construction of power lines had not been completed by 1930—to the best of our knowledge—and only one electric well was recorded there that year. In 1934, however, an internal census conducted in the colonies by the Farmers' Federation showed that electricity had become the main motive force for pumps in the Sharon and Samaria regions, as thirty-eight of forty-two pumps enumerated there (90 percent) were electric. In the colonies of Judea, in contrast, only 123 of 401 wells (31 percent) were powered by electricity that year. Thus, the trend is one of use of electricity in new planting areas and continued preference for internal combustion motors in old regions.[23]

Why were piston pumps more common in veteran Jewish citriculture? It turns out that both the piston pump and the internal combustion motor that powered it had several distinct advantages over the horizontal centrifugal pump and electric propulsion. The utilization rate (a measure of operating efficiency) of the piston pump was an excellent 80–85 percent, as against only 65–73 percent for centrifugal pumps. Thus, water was cheaper when delivered by piston pumps than by horizontal centrifugal pumps. Furthermore, the piston pump delivered the same quantity of water to any elevation needed, a considerable advantage in view of the depth of the water table at the time. The action of the horizontal centrifugal pump, in contrast, had to be calculated in accordance with requisite amount of lifting. Any change in this indicator was seriously detrimental to the pump's operating efficiency, and any increase in the extent of elevation steeply reduced the quantity of water delivered by the pump. Thus, for most growers, the use of a horizontal centrifugal pump led inevitably to a decline in pump output and an increase in the cost of pump operation.

Additionally, the horizontal centrifugal pump was a relatively new technological import in the early 1930s, and this led to difficulties in maintenance and, especially, repairs. Even the theory of "how a centrifuge works," the irrigation engineer M. Seitz wrote, was not clear at the time; from the standpoint of the experts in Palestine, it was still vague in several respects. Since the very process of recognizing and adopting this technical innovation was in its infancy, the horizontal centrifugal pump was not immediately and wholeheartedly accepted.

Moreover, the internal combustion motor had two important advantages in comparison to the electric motor. The first and decisive one was its low cost of operation relative to the electric motor, due to the very low price of fuel (kerosene or residual fuel oil) and its very high utilization rate. The electric motor, in contrast, was less efficient and drew on a very expensive source of power. The second advantage was the limited spread of the electricity grid, which ruled out easy connection of electric motors to the national system.[24]

Thus, the technological change embodied in the horizontal centrifugal pump and electricity as a power source for pumps was accepted selectively in the old citriculture areas. The reasons for this are rather clear. The main consideration for most Jewish growers certainly seems to have been economic. Electricity cost more, both in kW/hr and in the utilization rate of the motor, and was not even available everywhere. Although the piston pump entailed a large installation investment and had additional drawbacks, it was more efficient than the new centrifugal pump and the way it operated (delivery of a fixed quantity of water irrespective of the lifting height) met most growers' needs. The piston pump was also less expensive and more familiar. Furthermore, piston pumps had been widely diffused in old orchards before the country had a well-developed electrical system. This created a "tradition" of using these pumps—one that was reinforced by the economic advantages of the supply of repair and maintenance services, lower cost of the knowledge required to become familiar with the pump and its operation, and, above all, the fact that the investment in piston pumps had already been made.

Thus, the majority of Jewish growers, nearly all of whom in the late 1920s were farmers in veteran colonies, did not treat "modernization" as an imperative that must be embraced irrespective of its economic price. Experts such as Ryerson touted modernization as a model for progress, greater efficiency, and exceptionally high profitability in a developing country such as Palestine. However, local conditions played a decisive role in decisions for and against the adoption of technological innovation. Ryerson's proposals did more to set up an agrotechnical ideal, toward which one may aspire, than to meet the veteran citrus growers' needs. The drill pumps that Ryerson proposed as an alternative to suction-force pumps were not yet a practical alternative because the water table was close enough to the surface for the existing pumps—and, in some cases, close enough to flow in response to the urgings of the traditional waterwheel. The method of sinking wells also continued to be excavation and not drilling, since manual labor for excavation was cheap and there was no need to use the type of pump that a drilled well entailed. Under these circumstances, the California model was only partly adopted and, in many cases, attracted no response whatsoever when proposed. However, the tendency in the new planting areas of the Sharon and Samaria—especially toward the mid-1930s—was to adopt the electric drill pump.

The crucial difference between the two methods, of course, was in the cost of building the pumping system (well, pool, and pump). As a rule, Yakhin—and growers in the other new citriculture areas—wished to establish central water facilities that would be run mostly by drill pumps and powered entirely by electricity. A Yakhin promotional booklet, distributed in 1939 to potential investors, explained the method that the Histadrut-owned company had adopted: "Arranging [one's own] well is worthwhile only for an orchard floor no smaller than 100–150 dunams. In smaller areas, a cooperative well should be built or water should be purchased from a nearby well owned by a private individual or the colony itself." This is how the P£ 6,000 pumping system in Ramat Hasharon, dedicated on June 25, 1929, were installed: A well was sunk to a depth of forty-two meters, a 105-horsepower electric motor to power centrifugal pumps was installed, a seven hundred cubic meters pool was built, and six thousand meters of tubing were laid. Until the decision to establish the water system was made, Ramat Hasharon had only ten dunams of citrus orchards. By late 1929, there were 410 dunams. In 1933, the farmers' newspaper *Bustenai* reported that Ramat Hasharon had two thousand dunams of citrus, made up of small parcels of ten to twenty dunams apiece. The parcels were irrigated by a central water system that delivered more than cubic meters per hour.[25]

Interestingly, these findings about pumping systems in Jewish citriculture correspond largely (and after the fact, unsurprisingly) to models of technological adoption and diffusion and similar historical examples. In regard to models of technological adoption, Mansfield (1961) and many successors showed that adoption—and subsequently diffusion—takes place insofar as the profitability of its use rises and its investment cost falls. This model corresponds to our foregoing account about expected profitability from the use of efficient pumping systems in new groves and the diminishing investment cost as the method was applied over a larger area of land. For the opposite reasons most veteran growers refused to install new (even though, efficient) pumping systems. As for historical examples, Nathan Rosenberg showed that American manufacturing did not adopt the steam engine in one go, even though this invention was well known, because a handier and less costly source of motive power, water, was available. Only in locations where sources of water power were (or became) scarce relative to sources of fuel was the steam engine adopted.[26]

Pumping Systems in the 1930s

During the 1930s, it became increasingly necessary to deepen the wells as large-scale pumping caused the water table to decline and as the steady expansion and maturing of orchard area created a need to find abundant water sources. According to the growers' calculations, a mature orchard needed eight hundred cubic meters of irrigation water per dunam per year on average—four times as much as needed to irrigate the same orchard floor in the first years after planting.

To make the wells deeper, modern technologies of deep drilling and pumping had to be used. The water elicited by deep drilling was superior in quantity and quality to that brought up in ordinary wells and had the further advantage of being potable. This evidently made it more cost-efficient for several growers to establish central pumping and irrigation systems cooperatively than for each grower to install these systems himself. What is more, if this method also provided the colony with drinking water, the individual grower's share in the cost of installing and operating the central water facility and its adjunct, the efficient irrigation system, became even lower. Indeed, the sources point to a tendency to associate in the establishment of a medium-sized water company that would deliver water to a specific area.

In January 1934, the Agricultural Committee of Petah Tikva (the farmers' local representative agency) held a debate about the falling water table and the dwindling output of the wells and resolved to encourage growers to move over from private wells to a deep central well. A committee to plan out the initial project was set up and it was agreed that all growers would participate in its funding. Nine months later, the committee reported that measurements for the deep well had begun, that most of the existing wells remained in grave condition, their output was falling steadily, and there was no point in digging them any deeper. The goal was to sink a cooperative well to a depth of three hundred meters.

However, the predisposition to technological conservatism had by no means disappeared in Petah Tikva, the "mother of settlements." Ahead of its sixtieth anniversary (1938), the town had 116 private pumping systems, of which sixty (52 percent) were still powered by internal combustion motors and fifty-six (48 percent) had electric pumps. Just the same, this marked a considerable change from the situation in 1927, when few water facilities in Petah Tikva had electric power.

In the mid-1930s, for the average water table of the time—twenty-five to thirty meters—it was cheaper to build a pumping system for a drilled well at P£ 70–95 than to excavate a well and install a suction-force pump.

Jewish citriculture in Palestine had come a long way from the traditional well and waterwheel. In the late 1930s, even veteran colonies had begun to forsake technologies that Jewish citriculture had been practicing since the late Ottoman period, for example, piston pumps and residual fuel oil propulsion. However, the adoption of technological innovations in well and pumping facility design was not universal; in the late 1930s one could still find pre-Mandate methods and practices in old citriculture areas.

Two main factors seem to have affected the adoption or rejection new pumping system technologies: the depth of the water table and the cost of building and operating the system. In the latter respect, the price of electricity was crucial. Furthermore, the extent of adoption of new technologies was different in the orchards of the veteran fruit-growing colonies (mainly in the Judea area) than in the planting areas of the Sharon and Samaria. In Judea, citrus areas were expanded by veteran growers or their offspring, who tended to plant and cultivate in accordance with methods that they had known "since time immemorial." Moreover, the adoption of new technologies in the construction of pumping systems—foremost the drill pump and electrical power—required cooperative action on the part of several growers to lower the cost of installation and use to each investor. However, veteran growers did not forfeit their independence easily. As long their private wells delivered enough flow at a total cost that allowed them not to associate with others, they maintained the pumping system structure that they had been using since the beginning of the Mandate period: an excavated well and a piston pump running on kerosene or residual fuel oil. However, in the absence of accurate data about the regular operating cost of residual fuel oil and electricity, it is hard to know how willing veteran growers were to absorb losses in order to maintain their independence in water supply. On the surface, it seems that they tended to go over to cooperative pumping and adopt technological innovation only when they verged on a crisis.

In the new planting areas, in contrast, the new technologies were adopted vigorously. In those regions there was a prior intent, at the planting phase, to minimize production costs by adopting new technology-intensive cultivation methods that required little manual labor. New methods that required less daily labor and exploited scale economies were adopted in pumping systems, too.[27]

Irrigation System

The irrigation system was the second component of the composite water-delivery system. As explained above, the traditional irrigation system in Arab citriculture was comprised of a pool, ditches leading from the pool, and basins around the trees, to which the water eventually was delivered. Jewish growers in the late Ottoman period improved this system in several ways, for example, by sealing the pool and lining the ditches with concrete. This prevented water loss due to seepage from unlined ditches. The introduction of mechanical pumping at the wellhead made it possible—theoretically—to deliver water through tubes to the highest elevation in the orchard, whence it would flow in ditches to the basins surrounding the trees.

In the late 1920s, irrigation was done by means of manually hoed basins around the trees. In the best-cultivated orchards, a ring of earth was prepared closer to the trunk in order to keep the trunk dry during irrigation and, thereby, to reduce the risk of gummosis. Water was distributed in the orchard through imported metal tubes that, in most cases, received water directly from

FIGURE 5.3. Manual irrigation (Jewish grove).

the pump and through a system of ditches—lined and unlined—that delivered water from spigots that were installed on the metal tubes to the basins. The tubes were laid atop the ground. This distribution method was expensive due to the cost of purchasing the tubes, wear and tear, and labor intensivity in cutting the tubes and assembling them to fit. Experts stressed that this method, based on manual labor, persisted only because labor was cheap; as demand for labor and standards of living rose, it would inevitably become much costlier. Therefore, they urged the growers to adopt the California irrigation method.[28]

The expression "California irrigation method," coined at the time to denote the irrigation system used in the California citrus industry, referred mainly to underground installation of irrigation tubes and distribution of water by means of spigots. A basic condition for the use of this method was generous tree spacing, at six to seven meters. This left room to bury the tubes and cultivate between the rows by means of animals or tractors. To irrigate the roots themselves, a "furrow system" was used instead of ditches. Spigots attached to the underground tubes brought water to the surface and discharged it into dirt furrows that were ploughed in the orchards by mechanical means. Water reached the roots by percolating from the furrows, obviating the need for basins. In botanical terms, this could occur because the roots branched out from under the trunks and spread in greater part to the space between the trees. Furrow irrigation would be cheaper, it was argued, because machinery would replace manual labor in doing the work.[29]

In the late 1920s, however, the California method did not take root in the old citriculture areas; irrigation systems remained basically the same as in the late Ottoman period. The most important element of the old method, manual irrigation by means of basins, resisted all change. Water was delivered to the basins as before, through ditches that in many orchards were simply etched into the ground. One important change that did occur was the installation of metal tubes to deliver water from the pump. This, however, came about due to the introduction of more efficient pumps; it had no decisive effect on the nature of the irrigation system as such.

Things were different in the new planting areas in the Sharon. There, as stated, many orchards were planted with the express intent of maximizing the integration of technological innovations. A good example was the Yakhin company. "When we set up the irrigation grid," Yakhin stated in a 1931 report, "we introduced the California irrigation method with cement tubes. It is advantageous in that it reduces labor expenses for irrigation, allows the entire area to be cultivated by animals or tractor, and makes the installation itself less costly." However, the California method was also introduced in new planting areas that were not cultivated by Yakhin.[30]

Another attempt to modernize orchard irrigation was the introduction of sprinkling. However, "[a]n experiment in (artificial) irrigation sprinkling," it was written, "shows that it is not cost-effective in orchards: the moisture

FIGURE 5.4. The "California irrigation system" in Yakhin grove, 1936.

caused mold to grow on the leaves, grass between the trees proliferated, moving around the machinery and the sprinklers was very bothersome, very high water pressure is needed, etc."[31] In brief, the disadvantages outweighed the advantages. Technological innovation in this matter was tested and found unsuitable to the conditions of Palestine in the late 1920s.

Did the adoption of new technologies in orchard irrigation accelerate at the end of "big planting period?" In 1938, an interesting report published by the United States Department of Agriculture (USDA) about the Palestine citrus industry presented a concise description of the industry at large, including an account of its irrigation methods. This report speaks in a totally different tone than reports by various experts who visited the country previously. One reason, evidently, is the favorable disposition of its author, William Hazen, toward the Zionist Movement, as one may adduce from a personal report that he wrote in favor of the Zionist argument about Palestine's economic absorptive capacity. However, the contents of the citrus report show clearly that Hazen viewed the citriculture of Palestine through professional lenses. Furthermore, he wrote the report in his capacity as an economist with the USDA Economic Bureau. Instead of speaking condescendingly about the "backward" nature of local agriculture, the report expresses envy and concern about its successes in the international market. Although the report pointed out the deficiencies of citriculture in Palestine, it focused mainly on its strengths, in preparation for rising competition (expected and existing) between Palestine citriculture and its American counterpart.

The comments in the report on citrus irrigation in Palestine lead to two main conclusions. First, the basin method was still predominant; only 2 percent of the planted area was irrigated using the "California" method. The second conclusion, related to the first, is that irrigation using the furrow method, which proposed to replace manual labor with machine labor, had been adopted in large orchards but not in small ones.[32]

This comes as no surprise; it is consistent with the theory in this matter that stresses the importance of farm-unit size as a "threshold" for the adoption of a mechanized technology. Foremost among the factors listed as stimuli for the adoption in a large farm unit is the ratio of cultivation cost, including mainly labor cost, to mechanization cost. Even a large farm cannot expand ad infinitum if manual labor is used because of the constraint of supervision. In other words, in a labor-abundant farm the larger the cultivated area, the more supervision is needed and the greater the labor wage outlay becomes. Thus, at a certain "threshold" size (which varies depending on the conditions of the locality and the ratio of labor cost to mechanization cost), the transition to mechanized cultivation becomes cost-effective. Two additional factors give a large farm unit an edge over a small farm in adopting a technological change: greater specialization, which permits more efficient use of the new technology; and, the larger the firm, the more easily it can absorb a temporary decrease in output during the transitional period and put the technological innovation to more intensive use.[33]

Thus, the conditions in Palestine forestalled the wholesale adoption of the California irrigation method. A source from 1937 traces the wide acceptance of basin irrigation to "the great density of orchards of the Palestine." The impoundment pool also remained in vogue, because "it lets [the grower] put the well to better use by pumping at night, too." Sprinkling, a technology-intensive irrigation method, failed to make extensive inroads in Jewish citriculture even in the late 1930s. "In a few places in the country they have begun to irrigate orchards in that manner in recent years. . . . Irrigating with sprinklers saves lots of manual labor but there is an added expense for the installation of the grid of pipes."[34] Alternately expressed, the savings on labor costs did not offset the high cost of the investment needed to install the innovation.

Thus, the irrigation system employed in the orchards of Palestine was basically unchanged during the Mandate era. Its main components—delivery of water to basins, transfer of water through tubes and ditches, the impoundment pool, and the liberal use of manual labor—were invoked at the end of the period as they were at its beginning. High land prices and an abundant labor supply gave growers an incentive to maximize the production of each dunam by planting densely. Dense planting, in turn, forced the growers to irrigate the orchard by hand, and the most efficient way to do this was the basin method. Concurrently, the installation of central pumping systems, the boosting of well outputs, and the introduction of more powerful pumps allowed growers to augment the irrigation system with tubing. Thus, insofar as technological innovations were introduced in orchard irrigation, they did not affect the main component, manual irrigation in basins. It was the economic calculus—production per land unit—that ultimately determined the extent of the adoption of a new technology. Therefore, the introduction of technological innovations in the pumping and irrigation systems of Jewish citriculture in the Mandate era did not follow a "linear progression." Many innovations were rejected outright or adopted only in part, whereas others were adopted en bloc. Apparently, then, it was the economic consideration that counted most in adoption or rejection of technological innovations in this pronouncedly entrepreneurial industry.

Propagation

In the Ottoman period, as noted above, Jewish citrus growers went over from rootstock propagation to seed propagation. After the seeds germinated, however, the growers carried on in the local Arab manner. Each grower set up a nursery of his or her own and, some time later, planted the grown rootstock in its permanent location in the new orchard and grafted a Shamouti orange or other citrus species or variety onto it. In the Mandate period, two main propagation issues were of concern to the growers: (1) Should they graft the scion onto the rootstock in the orchard or purchase grafted saplings from nurseries that specialized in this? (2) What rootstock should be chosen, sour orange or sweet lemon?

Where to Graft—Orchard or Nursery?

Until the late 1920s, when the "planting frenzy" began, Jewish growers continued to propagate and graft saplings in their own orchards. When the planting rush began, nurseries that sold pre-grafted saplings came into being. Experts judged this to be an important time-saver:

> If a seedling spends four months in a seed bed, twelve months in the [grower's] nursery, and twelve months in the orchard until grafting can take place, it will take twenty-eight months just to reach the grafting phase. People who buy grafted saplings shorten this time. . . . Thus, while the Californians and the South Africans get a case or two of fruit from each tree by the fifth year, we would be pleased to obtain such a yield after seven years![35]

This considerable advantage of nurseries, however, came with an obstacle: the high cost of nursery-grafted saplings. Even at the end of the period at issue, this was the main factor in Jewish growers' decisions to introduce or reject this method.

In this respect as in others, the experts' recommendation was not universally accepted and the modernistic California model was not adopted en bloc. Still, few growers performed the entire propagation process in their orchards. Seeds were grown to the sapling stage in specialized nurseries and not individually in each orchard, as had been the case. The next phase, grafting the saplings, continued to take place in the orchard and not in a nursery due to its high cost.[36]

Sour Orange or Sweet Lemon?

Pursuant to the recommendations of experts such as Katzparovsky, Zagorodsky, and Pascal, and in view of the late 1920s modernization trend, there was a tendency of sorts in 1927–1932 to replace the sweet lemon with the sour orange as the preferred rootstock for the Shamouti orange because the latter was more resistant to disease. Opponents of the changeover stressed the advantages of the sweet lemon, especially its early and abundant yield. These advantages, under Palestinian conditions, meant that the growers could recoup their investment more quickly than with sour orange rootstock. Thus, the preference of the sweet lemon was taken for granted until the beginning of mass planting in 1926 and remained valid for rank-and-file growers later as well. In the estimation of the USDA in 1938, 90 percent of orange orchards in Palestine used sweet lemon rootstock. Here again, it seems that economic calculus overruled the experts' views.[37]

Picking and Packing

If any single aspect of Jewish citriculture in Palestine best reflects the image of this industry as a private, competitive, middle-class capitalist enterprise, it is unquestionably the way the finished product was picked, packed, and marketed. These activities, more than any other, carried the imprint of individualism and profit motive that underlay the existence and praxis of Jewish citriculture. Before, during, and even at the end of the Mandate period, each grower employed groups of pickers and packers on his or her own. A small building in the orchard, usually made of concrete or brick (but in some cases a large tent or a wooden shed), served as a packing house. Most of the Palestine citrus harvest was sorted and packed in these small and numerous facilities. Only a small portion of the harvest made its way to central packing houses.

Two main economic factors explain the decentralized nature of Palestine Jewish citriculture in the interwar years: the high export prices obtained for citrus produce and the availability of cheap labor. The profitability of citriculture—obtained by using labor-intensive technology—seems to have made the investment in technical innovations unattractive. This also explains the growers' opposition to the establishment of a central organization that might have done much to improve the harvesting and packing methods (see chapter 9 for discussion).

The packing and marketing policies of most Jewish growers were shaped by an environment of easy and immediate profits. The aim was to bring as many packed cases as possible to market; the quality of the fruit was not always the main criterion. A central organization that wished to build a good reputation in export markets would place picking and packing under tough supervision. Just the same, picking and packing methods did improve in some ways during the Mandate era. As the following brief account shows, these improvements flowed from the growers' wish to lower the cost of these phases of their activity.

Traditional Harvesting

> They did the picking with pruning clippers. During the entire harvest, an Arab woman stood on the ladder—next to the picker—with a basket on her head, or a basket was suspended on the branches of the tree with a rope and a hook. The picker would carefully place the fruit in the basket. They used to do the picking with rough clippers that might damage the fruit, but they did it very carefully and very few oranges were damaged. The workers were unique people who specialized in picking; they were paid more than ordinary workers.[38]

Once the baskets were filled, women and children carried them on their heads or their backs to the packing shed on the orchard premises. Sometimes field baskets were hauled to the packing shed on stretchers.

Technological Innovations in the Picking Process

The original ladders, which had to be leaned against trees, were replaced by more efficient ladders with movable legs. Pointed clippers were replaced by rounded clippers, resulting in much less damage to fruit. Pickers were given burlap sacks in which they placed the harvested fruit.

As yields and orchard size increased in the late 1920s, many Jewish orchards replaced their wicker baskets with wooden field boxes. The box was placed next to the tree and the pickers placed the fruit in it directly, if they were working at the bottom portion of the tree, or moved the fruit to the box from the burlap sacks in which they had placed oranges that they picked from the upper branches. Wooden boxes had two advantages over wicker baskets: They held more fruit and they were easier to store in the packing shed. This innovation increased the capacity of the packing shed and greatly reduced the number of haulers needed to move picked fruit to it. However, since the field boxes were large and bulky, they were placed on stretchers for haulage

FIGURE 5.5. Picking at "modernized" Jewish grove, 1930s.

to the packing shed, and this entailed the employment of men. This made hauling at least twice as expensive as in orchards that stayed with the old methods (baskets on people's heads or on stretchers) and employed women and children, whose wages were half those of men. What is more, an additional investment was needed to expand the packing house to store the cases.

While some Arab orchards continued even in the interwar era to haul fruit to the packing sheds in the old ways, fruit in larger orchards was taken to packing facilities in labor-saving carts or decovilles. Toward the end of the period reviewed, another haulage device was introduced, the *hadovile*. The hadovile, a small cart mounted on a single rail—unlike the decoville, which moved on two rails—was especially suitable for densely planted orchards. Shown to growers as early as 1934, it began to be assimilated in 1936, apparently because the Arab uprising made cheap manpower much scarcer just as the new orchards reached bearing age. The hadovile could carry six field boxes and was easy to install. Propelled by only one worker, it could replace ten women and children haulers. It was so inexpensive to purchase and install that, according to contemporary sources, even owners of tiny orchards could afford it.

By the end of the period reviewed, traditional methods of harvesting citrus and moving it to packing facilities survived only in medium and small Arab orchards. In small and medium Jewish orchards and large Arab orchards, where manual labor was more costly, many technological innovations in harvesting were adopted. The decoville, in contrast, made inroads mainly in large Jewish orchards and less in smaller ones.

It is also noteworthy that the expansion of total orchard area, the use of Jewish labor in harvesting work, and the wish to bring down production costs seem to have eliminated the expert picker. The quality and quantity of harvests often declined due to the employment of unskilled labor in this endeavor. This was especially common among Jewish workers, who had greater professional mobility.[39]

Traditional Packing

Even in 1939, the prevalent method of packing fruit remained the traditional one. The work was done by hand, the packers sitting on mats that were strewn on the packing-house floor. On one side of the packing house, fruit brought in from the orchard was piled in a heap, next to which the packing group sat. The group was headed by a packer, to whom the others were subordinate: selectors (who sorted fruit by levels of quality), sorters (who sorted the types selected by size), wrappers, haulers, and the "carpenter," who closed the cases with wooden or metal bands. The dominant packing method was the local one, in which oranges were placed atop each other layer by layer. The fruit in each layer was of a different size and an effort made to maintain uniformity of size in each layer.

FIGURE 5.6. Traditional packing at an Arab-owned grove.

Innovations in Packing

In 1922, a group of entrepreneurs attempted to introduce the efficient, American-style packing house model in Petah Tikva.[40] They purchased the packing equipment in Florida and installed it in the "mother of settlements." The experiment failed quickly and ignominiously, for three main reasons. First, the growers distrusted the new facility and brought it only poor-quality fruit. Second, the machinery was suitable for round oranges but not for oval fruit such as the Shamouti. Third and most important, there seems to have been no incentive to try to tailor the mechanized system to the conditions of Palestine. The growers' suspicion of any move toward centralization and the strong profitability of the industry seem to be at fault for the nonintroduction of this technical enhancement. As planted area and harvests increased in the late 1920s and early 1930s, large grower organizations such as Pardess and Jaffa Citrus Growers Exchange supported and initiated several attempts to establish mechanized packing facilities. However, these ventures, which were part of an attempt to subordinate individual growers to a cartel (an aspect that we discuss below), did not work out well. At the end of the period, only 2 percent of the total harvest was packed in mechanized facilities.

M. Goldberg of Petah Tikva introduced two innovations for small and medium Jewish orchards citriculture in 1933–1934. The first was a sorting table that allowed sorters to do all their work sitting down. After the selectors separated the fruit by levels of quality, the fruit rolled down a gradient between rods of steadily increasing width and fell into containers under the table in accordance with their size. The table made its debut at Pascal's orchard in Petah Tikva in March 1933. It was not cheap, at P£ 65 in 1938, and it was so bulky that the packing shed in the orchard had to be expanded to accommodate it. Furthermore, it required the use of field boxes, an innovation had not been adopted in all orchards (especially Arab ones, but also in Jewish-owned medium and small orchards). Accordingly, this innovation did not take root in Jewish citriculture.

The inventor tried to improve and adjust his invention through "learning by doing," bringing the table closer to the ground and adjusting it to the widely accepted selection and packing method—sitting on the floor. Thus, even small growers could avoid the cost of field boxes and make their packing method somewhat more efficient. However, there is no evidence that Jewish citriculture accepted the innovation even in its improved form. It did not help him that it appeared at a time of falling citrus revenues and attempts to contain costs.

Our sources show that the labor cost of picking and packing was largely unchanged between 1927 and 1938. The modest 6 percent increase in total picking and packing costs is attributable to an upturn in the prices of packing materials, which were imported. Materials accounted for much of the overall cost of picking and packing; their share actually rose from 55 percent of total picking and packing cost (per case) in 1927 to 58 percent in 1938.

In sum, there were differences between the adoption of technological innovations in picking and the adoption of similar innovations in packing. Quite a few innovations were implemented in picking, making this activity look much different at the end of the period discussed than at the beginning. The innovations at issue were rather inexpensive: shears, sacks, ladders, and to some extent field boxes. In large orchards, a decoville or a hadovile was also a cost-effective investment because it could slash labor costs. The technological improvements in picking also increased the quantities of marketable fruit.

Packing methods, in contrast, hardly changed: a small packing shed in almost every orchard; manual selecting, marking, and packing; and uncomfortable seating on the floor. The mechanization of packing entailed a large investment and the loss of marketing autonomy. High profitability and the abundance of cheap labor during most of the period impeded the adoption of technological innovations in packing.

By-Product Industries

Another tactic adopted by growers to enhance their profitability was the establishment of by-product industries that could use fruit that did not meet export standards and was not in local demand.[41] For such industries to exist, there had to be enough rejected fruit to serve as raw material and the endeavor had to assure its ability to generate a profit.

The first condition did not obtain in Palestine until the early 1930s. At that time, as production of fruit increased and as tariff barriers (camouflaged as health-protection regulations) were imposed on the export of citrus to Egypt—where the bulk of Palestine's low quality fruit had been sent—a surplus of unmarketed fruit came into being. Furthermore, over the years the Mandatory government had been placing the citrus industry, including the export phase, under increasingly tough regulation. From the 1936/37 season, for example, oranges and grapefruit that exceeded a certain size were not allowed for export. By limiting the amount of exported fruit the government hoped to assure a better price for it. The proportion of oversize fruit was estimated at 15 percent of the total crop. At the time, the share of the total annual harvest that was rejected due to damage was 10 percent and rising, as increasingly vigorous competition in European markets forced Jewish exporters to be more careful about the quality of fruit marketed and as government regulation of citrus exports became more stringent, as noted. Thus, 25 percent of the annual harvest in the second half of the 1930s could not be exported but was good enough for by-product industries. The swift growth in harvest quantities during those years, tracing to the maturation of trees planted in the late 1920s, led to a perceptible upturn in the absolute number of nonexportable cases. By the 1936/37 season, this number was estimated at between 2.5 and 3 million, definitely enough to sustain citrus by-product industries in Palestine.

The initiators of the by-product industries had ample and up-to-date knowledge. They were aware of quite a few citrus by-products, such as juices, syrups, frozen juices, canned fruit, alcohol, vinegar, lemon extract, industrial oil, soaps, color tints, powdered gelatin, jams, and ethereal oils (for liqueurs and perfumes). Rinds could be made into marmalade and other sweets, and the residues of the raw material could be used as cattle feed and fertilizer.

Several citrus by-product enterprises were established during the Mandate era. The most famous were Assis (formed in 1929 and still in business, with various ups and downs over the decades), Jaffora (est. 1935), and a few smaller enterprises. Just the same, in 1938 the by-product industries used only 100,000 to 150,000 cases of nonexportable fruit out of 2.5 to 3 million cases available. The proportion did not increase until the beginning of World War II. Profitability was low because the local market for these by-products was small and because manufacturers in Palestine had higher costs (wages, imported raw materials, haulage) than their rivals in Italy and Spain and due to high tariffs in export markets. Industry leaders asked the Mandatory government to help them by lowering the tariff burden in Great Britain and by providing convenient loans for investments in equipment and research. However, the government seems to have done nothing of substance in this regard during the period at issue.

SPECIES AND VARIETIES

In the Ottoman era, and more emphatically during the Mandate period as planted area burgeoned, the mainstay of Palestine citriculture was the Shamouti orange. Other citrus strains and varieties that are very familiar to us today were hardly accepted back then. They came on the scene belatedly and were mostly treated as side crops. A late 1936 census conducted by the Jewish Agency in the six main Jewish citrus areas of Palestine found that 82 percent of orchard floor in these areas was planted with Shamouti. Grapefruit accounted for 15 percent and all other varieties only 3 percent. The aforementioned USDA report for 1938 presents similar figures.[42]

Integration of grapefruit as the second most important species of citrus did not begin until the mid-1920s. According to the Jewish Agency census, 92 percent of grapefruit plantings in the six large citrus areas at the end of 1936 were planted between 1927 and 1936. The increase in grapefruit production was impressive indeed: in the 1928/29 season, Palestine exported about two thousand cases of grapefruit, whereas nine years later, in 1937/38, exports had increased by a factor of nine hundred to 1.8 million cases. Exports of oranges increased "only" 5.5 times over during those years, from 1.8 million cases to 9.6 million. However, the share of grapefruit in total citrus exports did not exceed 18 percent.

Other species of citrus—different varieties of oranges (Valencia, Washington), mandarins, citrons, etc.—were grown on a pilot basis only and did not attain commercial magnitudes. The only exception was the lemon, which was grown in somewhat larger dunamage at the end of the period reviewed than at the beginning. Even in the 1937/38 season, however, lemon production did not exceed 120,000 cases, and only two-thirds of the quantity was exportable.

Why did Palestine citriculture fixate on only one variety, the Shamouti orange? The main reason was the uniqueness and quality of the Shamouti as against the other citrus varieties and species in the market. It is this that allowed the Shamouti to compete very successfully with the other types of citrus in European markets. The Shamouti excelled in flavor, aroma, shape, resistance to damage in transit, and resistance to pests and diseases. Furthermore, the climatic and soil conditions in Palestine created unique circumstances that could not be emulated in other countries. This gave Palestine a monopoly in the production of the Shamouti—a situation that, unsurprisingly, the industry strived to maintain.[43] Accordingly, whenever an attempt was made to introduce a new species or variety of citrus, the question of its expected profitability was the decisive one for the grower.

Grapefruit is a case in point. Attempts to grow grapefruit in Palestine, such as that of the Palestine agronomist Abraham Brill back in 1913, after he visited the United States to bring saplings and scions of the Marsh Seedless grapefruit from Florida, failed. Palestine citriculture did not accept the grapefruit at the time. Only when trial shipments to Britain began in 1924–1928, proving that Palestine grapefruit was in strong demand in this market and, consequently, profitable, did extensive planting of grapefruit in Palestine commence.

Most grapefruit grown in Palestine during the Mandate era were of two varieties, Marsh Seedless and Duncan. Marsh Seedless was the more dominant; it was sweeter and juicier than the Duncan and, in direct contrast to the Duncan, was almost totally seedless. Grapefruit trees thrived in the climate of Palestine, gave a larger yield than the orange, and did so earlier—in the third year. The grapefruit also took well to the heavy soil in the Jordan and Jezreel valleys, where 90–95 percent of citrus area was planted with grapefruit. The fruit ripened by October in these hot regions and in November on the coastal plain, thus extending the season and improving profitability.

The grapefruit was more attractive than the Shamouti in terms of profitability, especially at the beginning of large-scale planting of grapefruit in Palestine. However, even though the grapefruit industry remained profitable until 1936, despite some decrease, grapefruit did not dislodge the Shamouti from its primacy in Palestine citriculture. There were several reasons for this: (1) Throughout the Mandate era, the grapefruit was considered a luxury and was therefore in limited demand. (2) Profitability did decline, for reasons including the Mandate government's decision to protect the produce of its colonies. In early 1933, tariffs were imposed on oranges and grapefruit at British ports. The tariff on oranges was set at 10 percent of produce value for the duration of the Palestine season and was higher in April–December only, the time when South African produce reached the British market. However, the tariff on grapefruit was high all year long at five shillings per hundredweight (about a case and a half). In 1933/34—the first full season after the introduction of the new tariff method—leaders of the industry claimed that the grapefruit tariff had slashed the profitability of this fruit by some 50 percent. (3) In view of the previous two factors, the future of the new industry was a matter of uncertainty. Warnings about uncontrolled expansion of the grapefruit industry dated from the very beginning of the large-scale planting period in the late 1920s. The auguries turned out to be correct several years later, when grapefruit profitability plummeted (as discussed in greater detail in the next section of this book).

By the end of the period, in the late 1930s, the perception of citrus monoculture had changed. As profitability declined and the supply of Shamouti oranges and grapefruit surged, people in the industry itself began to call for greater diversification. This, they thought, would extend the marketing season and, consequently, expand the possibilities of profit. The urgings for diversification focused on planting the Valencia orange, which ripens after the Shamouti and can be marketed until the middle of May. The Washington variety, which ripens about a month before the Shamouti, and mandarin varieties such as the Satsuma and the clementine, which also ripen before the Shamouti, were also proposed as ways to extend the season and diversify the industry. At this time, however, they were merely proposals or pilot plantings; they were not grown for export.[44]

Pests and Diseases

Pest and disease control is an important example of vigorous application of new technological know-how in the Mandate era.[45] The successful implementation of entomological and phytopathological knowledge in Jewish citriculture seems to have been inspired by all the general factors that spurred the introduction of modern technologies in the industry at large. The main factors, as noted above, include the Western orientation of the Jewish immigrant society that evolved in Mandatory Palestine, the availability of experts and others who disseminated the methods and knowledge among the growers, and, in the main, the very impressive cost-effectiveness of the innovations in terms of the added production attained.

Citrus diseases and pests were not serious problems during the era at issue. In retrospect, it seems as though growers did not need to combat them strenuously by introducing new technologies. However, contemporary sources indicate that even though citrus pests and diseases did little actual damage, they caused the growers much concern. Therefore, the lavish attention devoted to them in the literature and the professional press is unsurprising. Many growers evidently remembered the *phylloxera* epidemic that ravaged the fledgling Palestine viticulture industry and the 1915 locust plague that did much harm to the citrus industry as it existed then.

Furthermore, concern about a severe citrus epidemic seemed well placed in 1930 because a disease did make its appearance in the orchards of Palestine that year. Although it was labeled the "new disease" because it was unknown in Palestine theretofore, it turned out to be an old citrus disease called wood rot infection, which attacked the fiber of the sweet lemon rootstock, severely diminished the quality of the harvest, and afterward injured the tree itself. However, after no cure was found for several years, it transpired that even in areas where it had spread to as much as 90 percent of the orchard, few trees needed drastic care. The illness did little harm in most cases and the trees recovered unassisted within a few years. The "new disease" episode illustrates both sides of the citrus disease equation: the scanty damage such diseases actually caused during the period in review and the growers' continual concern about the menace.

Citrus Diseases

The main source of significant damage to growers was fruit rot. In seasons that were especially rainy or that gave forth exceptionally abundant harvests, for instance, 1936/37, up to 30 percent of the crop was damaged in this fashion. The rot, caused by a fungus, tended under extremely hot and humid conditions to spread rapidly to healthy fruit as well.

During the period discussed, growers tried to prevent fruit rot and its manifestations in various ways, such as strict care in harvesting and attempts to ship fruit in refrigerator vessels. These measures were of slight utility only. Not until the end of the period, in the 1937/38 and 1938/39 seasons, did the industry find an efficient (and inexpensive) method that, according to Isaac Rokach, "revolutionized shipping"—wrapping the fruit in paper doused with a chemical named Diphenyl. In this matter, the growers adopted knowledge that had existed since the early 1930s and had been piloted successfully in other locations and for other fruit. Pilot implementation in the 1937/38 and 1938/39 seasons proved that Diphenyl stanched the spread of fruit rot to healthy fruit during marine shipping and allowed the fruit to reach its destination in good condition and natural appearance. Experts claimed that the Diphenyl was not hazardous to health and did not permeate the fruit through the peel. Its odor dissipated either by itself, after the fruit was wrapped in it for a week, or after twenty-four hours if the paper was removed previously. It was very inexpensive to manufacture—about three Palestine mils per crate—and a machine that produced a million pieces of wrapping paper cost 150–200 Palestine pounds. The paper had a rather long shelf life: about a year without loss of effectiveness if wrapped in 2.5 kilogram parcels and properly stored. Thus, the advantages of Diphenyl were low cost (especially in comparison with refrigerated shipping and storage), the possibility of flexibility in shipping by eliminating concern about protracted storage, maintaining the appearance and flavor of the product, and no harm to consumers' health. Again, however, the use of Diphenyl was only experimental during this time.

Citrus Pests

One of the most important characteristics of citrus entomology in Palestine, as the famous Zionist zoologist and entomologist Simon Bodenheimer pointed out in *Hadar* in January 1938, was the little damage that most fruit pests caused there. "The growers will surely not agree with that statement," Bodenheimer continued, "but anyone who compares the situation here with that in other countries will find out right away that I am right." Bodenheimer cited two main reasons for this: the low proportion of citron and sour lemon rootstock, "favorite" hosts for many especially harmful pests, in the orchards of Palestine; and the powerful, dry, and hot *sharav* winds that blow through Palestine during the transitional seasons, which reduce the population of most pests to a minimum.

Most damage during the period discussed here was caused by several pests: (1) the Mediterranean fruit fly, which for the most part attacked ripe fruit of the end of the harvest season. The growers took no action to fight this pest; the best defense against it was to complete the shipping by the middle of May; (2) locusts, which did little harm during the Mandate era, in contrast to the major damage they caused in 1915; (3) various species of scales. The most pernicious pests for Palestine citrus during these years were of the last-mentioned type. Scales nestle in various parts of the tree, drink its sap, deprive the leaves and branches of moisture, and mar the appearance of the fruit so badly that it cannot be shipped. Some defenseless types of scales excrete a sweet fluid that causes a proliferation of fungus that blackens the fruit and the rest of the tree.

The growers fought the scales in three ways: spraying, fumigation, and biological warfare. The common ladybug, imported to Palestine in the Ottoman era by Baron de Rothschild's experts to fight the scales, minimized the population of the pest. Biological warfare against scales was even more widespread during the Mandate era. With funding from the agricultural committees in various colonies and from the Farmers' Federation, insect propagation facilities were established in Petah Tikva, Hadera, and Rehovot for the raising of various natural predators of scales, such as the scale ant-lion, a natural biological exterminator of the flour scale. Ladybugs were also propagated in these facilities, of course. Sources contemporary to the events show that the biological attackers were not very expensive to deploy, namely, about five hundred Palestine mils per dunam to deploy ladybugs in the late 1930s.

However, the war on scales was prosecuted mainly by fumigating the trees (e.g., in the campaign against the "nigra scale") and spraying. Private entrepreneurs quickly stepped in to provide both types of service. As against spraying by individual growers, as was done in the early Mandate era, fumigation against the "nigra scale" was performed by the government, which had made it compulsory. In the early 1930s, in contrast, agricultural extermination became a source of livelihood for professionals. Spraying and fumigation contractors did the work for the growers with greater expertise and at considerably lower cost. The cost of fumigation declined from two hundred Palestine mils per tree at the beginning of the period to thirty-five mils at the end. Now farmers could fumigate not only against the black scale—for which the government continued to pay at the end of this period—but also against the red scale and the hemispherical scale, since the cost of this measure had come down considerably.

The Government

In pest and disease control, as in other aspects of citriculture (as discussed at greater length in the chapters below), the Mandatory government intervened only gradually. It attempted, without great success, to stanch the spread of various pests from affected parts of the country to unaffected areas. In the late 1930s, it established an inspection service at the ports to combat the intrusion of pests and diseases borne on imported plants, set up several exhibition stations to demonstrate sound growing procedures and action against citrus pests and diseases, and launched a campaign against locusts in 1929 and 1930. As the industry grew, the government set standards for the inspection of nurseries and the diseases that they carried. The government also developed several experiment and exhibit stations of its own in Jericho, Sarafand, Acre, and other locations. After they did not prove effective, some were shut down and it was decided to subsidize the Zionist agricultural research station in Rehovot for citrus research provided that the findings be presented to the Arab sector as well. Thus, as the industry developed and the Mandatory regime became institutionalized, the government increasingly adopted the role that the pre-Keynesian liberal doctrine reserved for it: assisting the private sector without intervening in its actions excessively.

The Experiment Station

The Agricultural Experiment Station was established by the Zionist Executive in Tel Aviv in 1922. At first, citriculture was not an important part of the station's agenda, for two reasons: a scanty budget and the focusing of attention on the Zionist Labor Movement's cooperative and collective settlement enterprise, which specialized in dry farming. At that time, private growers obtained service mainly from ICA experts, some of whom were themselves growers. When the British Mandate era began, they could still consult with these experts and with Baron de Rothschild's erstwhile "gardeners." A noteworthy enterprise at that time was the Mikve Yisrael experiment station, which performed comprehensive and multifaceted citrus research throughout the period in review, with permanent funding from and active participation of the Farmers' Federation and its members. Another important source of knowledge for the growers was the professional literature that they amassed; this, for example, is how they familiarized themselves with California citriculture.

As the industry grew and the Zionist Movement farms began to take up citriculture, the experiment station in Tel Aviv changed its course of action. It encouraged the adoption of new imported technologies, took pains to diffuse them, and began to do research on citrus diseases and pests in order to solve them in original ways. In 1932, the station moved to its permanent location in Rehovot. When the move was made, the station received additional land, including five hundred dunams of orchards. In 1932, it had five departments that dealt with citrus issues, among other things: orchard selection, plant pathology, entomology, agricultural chemistry, and extension services.

Its research covered all phases of citriculture, from budding to nursery, young and mature orchards, and plant diseases. Experiments in shipping of fruit were conducted and recommendations for action were given in all cases. The entomology department, headed by Bodenheimer, performed various studies including a comprehensive investigation of "citrus pests in Palestine." Bodenheimer and his staff collaborated with growers in establishing small field laboratories, with funding from farmers' committees in locations such as Petah Tikva and Hadera, and used them for research on the citrus pests. It was then that the aforementioned insect propagation facilities were established in the citrus colonies. According to Bodenheimer, various entities, such as the Farmers' Federation and manufacturers of spraying oils, provided him and his department with funding for pesticide research. In 1934, the government gave the station a grant of P£ 3,000 to create experimental citrus farms. The Mandatory government established an inspection committee that received regular reports from Yitzhak Wilkansky, director of the station, and his team. Nevertheless, it is difficult to find specific areas in which the experiment station contributed significantly to the technological development of the citrus industry, probably because the facility went into action belatedly. It seems that it was the Mikve Yisrael station, and not the experiment station, that was more useful in fields such as irrigation quantities and timing, amount of cultivation, and correct fertilization.

In sum, much progress was made in adopting technological knowledge that pertained to citrus pests and diseases. The process began in the era of the ICA experts but gathered momentum during the Mandate era, when a larger number of players dealt in the field—the government, the experiment stations of Mikve Yisrael and the Zionist Organization (subsequently, the Jewish Agency), and the growers themselves. Two indispensable components made this development possible: low cost and a cultural predisposition to accept and adopt modern know-how. Here, in contrast to other aspects of citriculture, cheap labor and old methods were not alternatives. Thus, the process of technological adoption was rather inclusive.

INSIDE THE "BLACK BOX" OF PALESTINE JEWISH CITRICULTURE

The following discussion examines the extent to which technological innovations in Jewish citriculture during the years at issue helped to increase production per land unit (the dunam) and/or to cut production costs per production unit (the orange case) and land unit.

Table 5.2 illuminates the most important change that occurred between 1913 and 1937: an increase of hundreds of percent in the fifth-year yield for a relatively small 100 percent increase in investment cost. This explains the steep decline in cost per case between the eve of World War I and the eve of

World War II. Table 5.3 shows changes in factor costs and the total investment relative to the 1913 level (in constant prices).

Table 5.3 shows that the total cost of orchard investment in the first five years in Jewish citriculture doubled between 1913 and 1937. However, production in the fifth year leaped by 800 percent during this time, giving evidence of the technological improvements that had been introduced in propagation and cultivation methods. The numbers point to a somewhat clashing trend in investment cost between 1929 and 1937. Here it should be borne in mind that the costs in the 1929(2) column were those incurred by Yakhin, which, as we have seen, did not go out of its way to report full costs. The price of land shown in Yakhin's report pertains to new planting areas, which in column 1929(1) pertains to a veteran grower in a large and long-established colony, and the prices in 1937 are averages for the industry at large, in which the differences between old and new citriculture had been smoothed in the meantime. Thus, the price of land seems to have risen considerably in the Sharon citrus areas relative to 1929 but at a much lower rate relative to the old growing areas.

The overall picture is one of considerable increase in land prices, stability if not decrease in cost of the irrigation system—evidently due to the replacement of local wells in individual orchards with central pumping facilities, which were usually electric powered—an increase in labor costs, and an increase in the cost of the capital component, chiefly for manure and fertilizers. However, as stated, the leap in productivity between 1913 and 1937 points to the technological improvements in Jewish citriculture during the period at issue in this book. Thus, they give evidence of what happened inside the "black box," as Nathan Rosenberg calls it—the process that we have tried to describe and explain thus far. As noted, the improvements slashed the production cost per case of oranges by 80 percent between 1913 and 1937.

It is clear that the Jewish growers managed to move up the first significant harvest. Did they also step up production per dunam in mature orchards? At first glance, output per dunam decreased during the period at issue. The "Fruit Committee" estimated an average harvest of about one hundred export cases per dunam in 1927. Other sources speak about much larger harvests, 120–180 cases of export fruit per dunam. At the end of the period, however, the accepted estimate of production in an "average" orchard was no more than eighty export cases per dunam.[46] Those of the "nostalgic" persuasion among the old farmer class credited the strong production of producing orchards in the years up to the 1930s to the quality of their cultivation and care, and complained about the decreases that had occurred in the harvests and quality of cultivation in the new citriculture. They overlooked the important fact that before 1927 most producing orchards had been planted before 1914. Indeed, the productivity of an

TABLE 5.2
Five-Year Investment per Jewish-Owned Orange Orchard Dunam and Production in Fifth Year, 1913, 1929, 1937
(P£, 1936 prices)

Component	*1913*	*1929(1)*	*1929(2)*	*1937*
Land	7.00	16.20	9.26	19.28
Well and irrigation system	10.85	21.41	9.26	11.57
Labor	10.89	21.90	33.33	20.25
Capital	8.22	21.17	21.30	25.07
Interest	13.03	29.63	23.15	24.11
Inspection and supervision, taxes	—	4.63	Included in capital	Included in capital
Total	49.99	114.94	96.30	100.28
Cases per dunam in fifth year	5	20	—	45
Cost per case in fifth year (before picking and packing)	P£ 9.99	P£ 5.7	—	P£ 2.2

Sources:

1913—Meir Apfelbaum, "Citrus Growing in Palestine," *Haaklai* 1 (1912/13): 116–18 (Hebrew).

1929(1)—Melekh Zagorodsky, *The Citrus Book: Manual for Orange Growers, Workers, Experts, and Nurseries Owners* (Tel Aviv: Hahaklai, 1929), 211–16 (Hebrew). This is a budget for a fifty-dunam orchard in one of the old colonies, cultivated by means of pre–World War I methods.

1929(2)—Zagorodsky, *Citrus Book*, B, 220–21; Yakhin, *Marketing Bulletin for 1929*, "Orange Orchard Expenditure and Income" (pages unnumbered). This is a budget of the Histadrut-owned Yakhin company. All labor is Jewish and the cultivation methods are mechanized and more efficient than in the 1929(1) orchard. The extensive nature of planting and cultivation brought scale economies, thereby explaining the small cost of implementing the well and irrigation system, relative to the 1929(1) budget. The price of land seems too low; the Yakhin proposal itself notes that it could come to as much as P£ 25 per dunam.

1937—Shmuel Yedidyah, *Citrus Growing* (Tel Aviv: Hasadeh, 1937), 232–37 (Hebrew). The author presents data for "minimum," "optimum," and "maximum" prices. I chose the "optimum" figures for the table. Therefore, the table adds interest statistics that Yedidya does not present but that, as he notes, may come to "3–5 Palestine pounds per dunam per year" (ibid., 236), a computation that is somewhat more "optimistic" than mine. Yedidya's reckoning of 40–50 cases per dunam in years 4–5 (ibid., 194–96, 232) is corroborated by other sources. See Moshe Smilansky, "Our Agricultural Year," *Bustenai* 8:21–22 (Sept. 14, 1936): 17–20; Yakhin, *Marketing Bulletin for 1929*; Ludwig Gruenbaum, *Output and Structure of Expenditure in the Jewish Citrus Industry* [Unpublished manuscript] (Jerusalem, 1939), 6 (Hebrew).

TABLE 5.3
Change in Cost of Cultivating and Preparing One Dunam of Jewish-Owned Orange Orchard in First Five Years—1913 as Base Year, in Comparison with 1929 and 1937 (in 1936 prices and percent)

Component	*1929(1)*	*1929(2)*	*1937*
Land	(+) 131	(+) 32	(+) 175
Well and irrigation system	(+) 97	(–) 15	(+) 6
Labor	(+) 101	(+) 206	(+) 86
Capital	(+) 158	(+) 159	(+) 205
Total (including interest)	(+) 130	(+) 93	(+) 101
Cases per dunam in fifth year	(+) 300	—	(+) 800
Cost per case in fifth year	(–) 43	—	(–) 78

Source: Table 5.2.

orchard, assuming that it is cultivated satisfactorily, is above all a function of its age (even though a large harvest also requires larger capital and labor inputs). The Table 5.4 shows the ratio of age to production per dunam.

In 1927, most orchards in Palestine were producing orchards that were thirteen years old or older. This is because almost no new orchards had been planted between the beginning of World War I (1914) and 1923/24. In early 1937, in contrast, only 15 percent of Jewish-owned orchard area was planted with orchards ten years old or older. Some 42 percent of orchards that year were five years old or younger, too young to produce fruit.[47] Thus, the average production per dunam of mature Jewish orchards decreased in the late 1930s not because the quality of cultivation and care had diminished but because the extensive planting that began in the late 1920s had driven down the average age.

However, as shown above, most sources indicate that even if the output of a producing orchard at full maturity was not strongly dependent on the cultivation technology, the new technologies that the Jewish growers had introduced had a decisive effect on the *beginning* of significant yields. Fertilization, a two-year decrease in the period from seeding to grafting, and some increase in spacing resulted in production of between forty and fifty cases per dunam by the end of the fifth year. During the Ottoman era, in contrast, a five-year old orchard produced only five export cases per dunam.

As for Arab citriculture, in contrast, almost all sources point to a significant increase in production per dunam of fully mature producing orchards at the end of the period under discussion relative to the scanty output of the traditional Arab orchard. Indeed, statistics on Arab citriculture—some of which

TABLE 5.4
Age/Output per Dunam of Orange Orchard

Orchard age	*Export cases per dunam*
7–8 years	70
9–10 years	90
11–15 years	110
15+	130

Source: B. Sheinfeld, "Fundamentals of Orchard Evaluation," *Hadar* 6:1 (1937): 10–11.

are cited below—show that this sector of the industry changed during the Mandate era by moving toward the adoption of cultivation methods that the Jewish growers had begun to use. This adoption, which Jewish citriculture had carried out during the Ottoman era, caused the output of producing orchards to increase in a way that was not a function of orchard age. Thus, by implication, the intersectoral disparity in per-dunam production of producing orchards narrowed during the Mandate era.

Table 5.5 compares the cost per export case in 1926 and at the end of the period. The 1926 data pertain to an export case in the old citriculture, which reflects its dominance at the time; the second source (1938 data) is an average for Jewish citriculture at large, including both its new and its old segments.

The general picture shown in the table is an 8 percent increase in FOB cost per export case in the twelve years between the onset of the "planting frenzy" and the citriculture crisis of the late 1930s. Most of the upturn traces to increases in the price of packing materials, the cost of picking and packing, and capital recovery per case. The cost of haulage to port did not increase significantly. This evidently attests to a dual trend: an improvement in overland transport routes between the citrus areas (old and new) and Jaffa, and the introduction, by the end of the period, of rail service for haulage of most Jewish fruit from distant areas such as Rehovot and Nes Tsiyyona to Haifa port. Haulage of citrus to port usually took twenty-four hours and rail fares were kept high for most of the period. Thus, it is clear that the upturn in cost of overland haulage traces to longer distances to port and the cost of haulage by rail.[48] Nevertheless, as Table 5.5 shows, overland haulage remained a minor factor in the total cost per export case, at 12 percent in both 1926 and 1938.

Most of the decrease in cost per case derived from a reduction in total cultivation cost. This reflects growers' attempts to make do with less cultivation, mainly in old orchards, and to employ greater mechanization in current cultivation, especially in new orchards. In contrast, the increase in cost of

TABLE 5.5
Current Costs per Case of Oranges, Jewish Grove, 1926, 1938

	1926		*1938*		
Component	*Palestine mils*	*In 1936 prices*	*Palestine mils*	*In 1936 prices*	*Change in 1936 prices, percent*
Cultivation	114.8	86.8	85.5	84.2	(–) 2.9
Picking and packing	56.4	42.6	46.0	45.2	(+) 6.1
Materials	69.9	52.8	64.7	63.7	(+) 20.6
Haulage to port	41.8	31.6	32.7	32.2	(+) 1.8
Taxes and misc.	39.5	29.9	24.5	24.1	(–) 19.3
4% interest on foregoing	12.9	9.7	10.1	9.9	
Capital recovery, 8% for 30 years	108.1	81.7	104.4	102.8	
Total	443.4	335.1	367.9	362.1	(+) 8.0

Sources:

For 1926—Melekh Zagorodsky, *The Citrus Book: Manual for Orange Growers, Workers, Experts, and Nurseries Owners*, B (Tel Aviv: Hahaklai, 1929), 227 (Hebrew), at 125 cases per dunam. Zagorodsky does not state whether the figures for that year are in Egyptian pounds or Palestine pounds. Presumably the currency is the Egyptian pound, since he states explicitly that he obtained the data from growers in 1926 (before the Palestine pound was adopted). Thus, I converted his figures into Palestine mils at the official exchange rate of 1.02564 Palestine pounds to the Egyptian pound.

For 1938—Harry Viteles, "Survey of the Citrus Industry in Palestine," *Hadar* 11:1, 2–3 (1938): 22–23, at 100 cases per dunam. These sources do not calculate interest on current expenses (and such interest data as exist are severely incomplete) and capital recovery. These were added in the table above.

picking and packing—activities that were much harder to mechanize and were naturally intensive in manual labor—apparently traces to an increase in labor wage and greater use of Jewish labor. Taxation, another minor factor in cost per case, decreased due to the aforementioned decrease in property tax on farmland.

Notably, however, the number of cases per dunam was different at each of the two points in time. The output of a mature orchard was calculated at 125 export cases per dunam in 1926 and one hundred cases in 1938. The difference reflects the disparity in the age of productive orchards between the two periods. Accordingly, the costs of labor and capital recovery (both of which computed on a per-dunam basis) were "spread" over a larger number of cases in the 1926 orchard than in the

1938 orchard, thus reducing the per-case cost. If we compute the cost of a case per 1926 orchard on the basis of one hundred export cases per dunam, we find that the total cost per case decreased by 4 percent between 1926 and 1938. In any event, this difference in the number of export cases and the resulting increase in cost per case accurately reflect the profitability of citriculture during the years at issue.

SIX

Technological Innovations in Arab Citriculture

THE DEFINITIVE STUDY about the history of Arab citriculture in the Ottoman and the Mandate eras has yet to be written. It is true that every study about the workings of the Palestinian-Arab economy at that time mentions the importance of citriculture for Arab economic growth in Palestine and notes the patterns of modernization that citriculture brought to traditional Arab agriculture.[1] This book, too, does not aim to study Arab citriculture in any comprehensive way. Nevertheless, insofar as the sources permit, it will examine Arab citriculture in the principal areas in which it examines Jewish citriculture.

In 1914, as stated above, there were 30,000 dunams of citrus orchards in Palestine—7,500 Jewish-owned, 750 German-Templers-owned, and the remainder, 21,500 dunams, Arab-owned. These figures hardly changed between then and the end of World War I, and even several years later, in 1922, the total planted area was estimated at 29,000 dunams. Things began to change in 1923, as Jewish citriculture planting embarked on an uptrend, albeit on a small scale, whereas Arab citriculture showed no further planting until 1926 and subsequent years. The sources used in previous studies shed no light on the reasons for the lag in the Arab sector. Below, on the basis of new sources that we cite, we provide a partial explanation.

In any case, from 1926 on Arab citriculture and Jewish citriculture developed along very similar lines, although development in the Arab sector was slower. Arab growers planted less than Jewish planters each year, so that by the end of 1929 the Jewish planted area (46,000 dunams) surpassed the Arab area (41,000). The absolute disparity widened until 1934, by which time the Jews had 145,000 dunams of orchards and the Arabs 105,000. In 1935, Jewish planting activity began to slow considerably due to the steep decline in profitability, and in late 1936 planting in the Jewish sector came to a nearly total halt. In the Arab sector, in contrast, planting continued on a rather large scale in 1935–1936. Therefore, by the end of 1936 the gap had narrowed to 155,000 dunams of Jewish-owned orchards and 143,000 dunams of Arab-owned orchards.

A similar change took place in the extent of planting between the two subperiods: Planting in the Jewish sector accelerated from 9,750 dunams on annual average in 1926–1931 to 25,000 dunams on annual average in 1932–1934. Much the same occurred in the Arab sector, average annual planting accelerating from 5,400 dunams to 17,800 dunams in the respective subperiods.[2]

This points to a strong correlation between the extent and cycle of planting in the Jewish sector and the same parameters in the Arab sector. The reasons also seem to have been similar: strong profitability until 1926, which allowed capital to accumulate, and a capital inflow by means of Jewish immigration, which made the "planting frenzy" in the Jewish sector possible. Indeed, there is a correlation between the resources available to the Yishuv (which derived mainly from capital inflow) in 1926–1931 and 1932–1934 and the extent of Jewish citrus planting during the respective subperiods.

The Jewish capital inflow was evidently one of the main factors in the extent of Arab citrus planting. Some of this capital moved to the Arab agricultural sector by means of land purchases by Jews. In 1927–1931, Jews bought 155,000 dunams of land from Arab inhabitants of the country, and in the following five years, 1932–1936, the Jewish-held area nearly doubled to approximately 335,000 dunams. Thus, Jewish land acquisition was an important source of capital for Arab citrus growers. However, as stated, Arab growers continued to plant vigorously in 1935 and 1936, after planting in the Jewish sector slowed. The main reason was that Arab growers produced a case of export citrus at lower cost than the Jewish growers. Thus, the Arab sector of the industry remained profitable after the Jewish sector tumbled into a severe crisis. It seems, however, that Arab growers apparently had access to an additional source of capital other than the sale of land to Jews (which, as stated, continued even after the Jews stopped planting): profits that had accrued over the years from the activity of the industry itself.[3]

The prevailing view among scholars of the Arab economy during the period under discussion is that the citriculture was controlled mainly by the wealthy stratum, since this stratum owned most of the land. However, according to estimate made by Charles Kamen, large estate owners held only between 25 and 45 percent of the land on the coastal plain, where most citrus planting (Jewish and Arab) took place. Consequently, not only "deep pockets" but also small Arab landholders could venture into citriculture there.[4]

Below we cite several cases—mainly on the basis of archive sources but also on the basis of more detailed reports and surveys and growers' journals—

that illustrate the relationship between Jewish settlement activity, land purchases, and citriculture development, on the one hand, and the growth of Arab citriculture, on the other hand. We will also emphasize the technological aspects of Arab citriculture. Importantly, however, this information will constitute not a comprehensive study of the topic but rather a contribution to the issue from the perspective of the theme of this book.

YA'AKOV LUBMAN HAVIV

An important source in the survey to follow is a set of detailed reports from Ya'akov Lubman Haviv in 1932 to Moshe Shertok of the Jewish Agency Political Department about "dispossessed" Arabs (to use the term in vogue at the time).[5] Lubman Haviv was the son of one of the first inhabitants of Rishon Lezion, the surveyor and educator Mordechai Lubman, and was born in the moshava. After a nomadic period in Germany and Lithuania, Lubman Haviv returned to Palestine in 1920 and was hired by the Palestine Land Development Company as an assistant to Yehoshua Hankin. Lubman Haviv was fluent in the languages used in Palestine, well versed in the country's customs, and skilled in recording title to land.[6]

In the background of Lubman Haviv's reports were various government investigations and reports, prompted by the Arab uprising of 1929, about the causes and scale of the "dispossession" of Arab peasants. Lubman Haviv had a detailed list of villages and individuals who claimed "dispossessed" status. Accordingly, his survey, which attempted to provide a maximum of information about the previous and current economic situation of the claimants and their villages, was extremely detailed. His reports took a clear stance that corresponded closely to the official Zionist view of the utility that the Arab villagers had derived from the Jewish settlement enterprise. Thus, their Zionist bias is unquestionable. However, since they were secret and highly detailed, they contain much information that was meant for Shertok's eyes only. They also contain clear economic logic.

One may distill the following generalization from these reports: Until 1924, the Arab villagers were in a dire economic state. The joint method of land possession in the Arab village *(musha')* and the villagers' pressing debts to usurers thwarted any incentive to cultivate intensively and to invest. What is more, the villagers had no outside source of capital with which they could break the vicious cycle of *fallah* subsistence farming, which was unable to increase output per land unit beyond the basic needs of the peasant and his family.

Jewish immigration, especially after 1924, ruptured the equilibrium of this traditional agriculture. First, many villagers found an additional source of income by working for the Jews, for instance, in the citrus orchards of the moshavot and the large private plantations in the Judea and Sharon areas.

Second, many Arab landowners—large and small alike—sold some of their holdings to Jews and, after using the consideration to pay back their pressing debts, had enough capital left over to invest in intensive irrigated agriculture. When these effendis sold land to the Jews, they even compensated some of their sharecroppers, who invested this money, too, in irrigated agriculture. The most common types of irrigated agriculture were vegetables and citrus. Concurrently, villagers benefited from continued employment in the moshavot, the knowledge they acquired while working in the Jews' orchards, and the growing motor transport infrastructure, which reduced travel time to the urban centers significantly. The technology that they used, foremost in pumping and distribution of water, was modern, since such was the technology that they had learned from the Jews.

Lubman Haviv's reports also stressed the importance of private ownership of land in Arab villages as an incentive and a condition for the development of market-oriented agriculture. Therefore, even though the registration and mapping of land in Arab villages proceeded at a snail's pace in the early Mandate era, and even though the implementation of the 1928 Land Settlement Ordinance concerning registering of village land and transferring it from joint possession *(musha')* to private possession *(mafruz)* also moved slowly, the influence of these modernization processes on rural Arabs was evident in almost every line in Lubman Haviv's reports.[7] The possibility of establishing private landholdings created a strong incentive for the sale of the land even before the Mandate government's land settlement came to these villages. Importantly, too, Lubman Haviv 's reports show explicitly that many purchases from villagers were made not by Jews but by Arab investors from Jaffa and elsewhere. These investors, too, planted citrus orchards on their new holdings.

In the vicinity of *Rehovot*, *Nes Tsiyyona*, and *Rishon Lezion*, residents of several villages—'Agr, al-Na'ni, Zarnuqa, al-Maghar, al-Qubayba, and other villages in this area—sold land to inhabitants of Rehovot and Nes Tsiyyona. As noted above, Moshe Smilansky (along with Tovia Miller of Rehovot) was highly active in buying land from these villages for his moshava and for the Jewish National Fund. Zarnuqa, for example, "sold 6,000 dunams to the Jews" out of the thirteen thousand dunams in its possession. After this, according to Lubman Haviv, the village had "about 2,000 dunams of orchard, some of which are already yielding fruit and some are eight years old, and the land settlement was completed only three years ago. The same situation prevails in Sarafand al-Kharab, which sold half of its land, 5,000 dunams out of 10,000, and has about 2,000 dunams of orchard." Lubman Haviv also notes that "almost everyone in the [Sarafand al-Kharab]" had orchards, some of which were yielding fruit.

Some land of Kafr Saba village was sold to the ICA before World War I and the moshava of *Kfar Saba* was established on it. Other parcels were sold after the war to individual Jews and private companies and used for large-scale planting of orchards. Effendis gave some inhabitants of Kafr Saba unofficial compensation "for relinquishing their rights as de facto holders of the land or as sharecroppers." They used the money to plant orchards that ranged from twelve to two hundred dunams in area. Many villagers also worked in orchards of the surrounding areas.[8] Lubman Haviv underscored the importance of the new public transport for the vegetable growers among the villagers. "Connections with the moshavot and Jaffa and Tel Aviv are very convenient, and all the villagers use the automobiles of the moshavot. Also, their crops—vegetables, watermelons, etc.—are hauled only by the automobiles of the moshavot instead of camels."

Miska, north of Kfar Saba, did not develop as other villages did and may, in this context, serve as an example by negation of the utility that Lubman Haviv found in the Jewish moshavot as an incentive for the development of traditional Arab agriculture. Jewish private investors planted two large orchards near this village. One of them (est. 1927) was known as Kalmanya. It was named for the father of the owner, an important importer from Great Britain, Moshe Gardinger. It was 1,100 dunams in area and was managed by a veteran grower from Petah Tikva, Baruch Ben-Ezer. Gardinger himself lived in Tel Aviv and quickly became an important personality in the new private citriculture. The second orchard belonged to the Gan Chaim company, named for Chaim Weizmann.[9] The distance between Miska and the Jewish moshavot thwarted the village's development, Lubman Haviv claimed. No citrus area had yet been planted there; the village had unirrigated plantations only. However, "After another portion of the village is sold off and the boundary of the Jewish-owned area draws closer to them, and especially after the *musha'*, which accounts for more than half of the land the village has, is apportioned, the village will definitely become wealthier and its plantation areas will increase."

The land of the Bedouin tribe named *Abu Kishk* was north of the Yarkon River, abutting the Bachriya orchard. Half of the twenty thousand dunam area was sold. Approximately eight thousand dunams were acquired by Jews and three localities—the moshavot 'Ein Hai and Ramat Haim and Moshav Hadar—were established there. The remaining two thousand dunams were sold to Arab investors in Jaffa. The Land Settlement Committee, Lubman Haviv notes, had already visited the area and recorded title to the land in the name of the sheikh of the tribe and his brothers.

> Sheikh Shaqer and his brothers, who sold most of the *miri* land [land owned by the ruler but legally possessed by the peasant] that they owned, first paid off the oppressive large debts (about 10,000 Palestine pounds)—mortgages, etc.—and apart from that planted a 145–dunam orchard near their home, with a fifteen-horsepower motor and a six-inch pump. . . . Sheikh Shaqer and his brothers [also] planted two large vegetable gardens on the bank of the Yarkon.[10]

Quite a few studies have been written about the history of land acquisitions in *Al-Shaykh Muwanis*, the village where Tel Aviv University, the Tel Aviv fairgrounds, the Ramat Aviv neighborhood, and other properties stand today.[11] Here we concern ourselves only with the aspect of the development of Arab citriculture. "Before 1924," Lubman Haviv relates, "the village land was not totally cultivated." Farming was rudimentary and the villagers barely made ends meet. Only the Beidas family, an important landholder in the village, had a two hundred to three hundred dunam orchard near the Yarkon River, irrigated by means of a waterwheel. However, after the Fourth Aliya began and the moshavot Herzliyya and 'Ir Shalom (Ramat ha-Sharon) were established, about 1,500 dunams were sold to Jews. Furthermore, the land settlement was applied in the village. Those who sold their lands to Jews planted citrus orchards. Lubman Haviv listed ten different families that established orchards ranging from fifteen to 124 dunams apiece. Many also expanded their vegetable farms and "installed modern motors and pumps to pump water from the river." Several villagers planted orchards with earnings from the sale of vegetables to moshavot and Tel Aviv. "And all of them," Lubman Haviv writes with emphasis, "[did so] with modern machines and new irrigation facilities. They irrigate even the vegetable areas, which now amount to 2,000 dunams, by means of pumps and metal tubes. One hardly sees [pumping by] animal labor in the area any longer." As for transport, Lubman Haviv stressed that "automobiles" and trucks to Tel Aviv and the moshavot also served the farmers of Al-Shaykh Muwanis, "who once had to waste a whole day bringing produce to the city market."

Lubman Haviv 's review includes additional villages that bordered on Herzliyya, Ramat ha-Sharon, Ra'ananna, and settlement areas farther from Tel Aviv, both in the direction of Petah Tikva and as far north as Netanya. The findings were similar; they varied commensurate with the proximity of the village to a moshava or Jewish city and to transport routes.

THE CONTEMPORARY PRESS

"Jewish Colonization and the Fellah"

In this article, written as part of the official Zionist response to the allegations of dispossession that followed the 1929 violence, Moshe Smilansky attempted to demonstrate the contribution of the Zionist settlement enterprise to the development of the Arab peasant economy. Smilansky's survey drew a dichotomy between the Arab village, its inhabitants' way of life, and their culture as "primitive" and "slavish," and the Jewish settlement as "modern," successful, and "civilized." Pursuant to this argument, Smilansky tried to show that proximity to Jewish moshavot contributed to the economic growth of Arab villages that had this good fortune and that villages far from Jewish settlements remained backward. One may, it seems, accept the statistics without accepting their tone of voice and underlying worldview. Smilansky's remarks indicate that the situation in many Arab villages near moshavot—such as Zarnuqa and al-Qubayba near Rehovot, Sarafand and Bayt Dajan near Rishon Lezion, Rantiya, Kafr 'Ana, Salama, Saqiya, and Yehudiya near Petah Tikva—resembled the depiction in Lubman Haviv's reports. The sale of portions of village land to moshavot, wage agricultural labor, and the sale of services and goods to moshavot provided villagers with resources that they used to plant orchards and switch to intensive cultivation of vegetables. Overall, Smilansky counted some six thousand dunams of orchard area, much of which belonged to low-income villagers and not only to members of the wealthy class.[12]

Netanya

Reportage in a regular column in *Bustenai*, called "From the Moshavot," told the story of Netanya's economic success. Land acquisition was carried out through an Arab mediator from Tulkarm who, on one hand, was an activist in the Arab national movement, and on the other, engaged in land transactions with the Jews (the figure was probably Selim Ab'd al-Rachman from the Al-Haj Ibrahim family.[13]) The land of Netanya itself had belonged to Sheikh Salah Khamdan of Umm-Khalid (near Netanya). Sheikh Khamdan had amassed burdensome debts to creditors from Tulkarm. However, *Bustenai* reports, "[he] found relief and salvation by selling some of the land to the Jews: the debts were erased, a large orchard of hundreds of dunams was planted, and a modern house was built."[14]

Tulkarm

The journal *Mis'har ve-Ta'asiyah* (Trade and Industry) occasionally carried reports about events in the country's Arab sector. In 1932, there was a spree of reportage about new planting activity in the region of Qalqilya and Tulkarm, situated alongside the Jewish Sharon plain and to its east. The reports from Tulkarm are especially interesting: large citrus areas were being planted by wealthy investors from Nablus, the "metropolis" of the Tulkarm area. Ahmad Shak'a, Taher al-Masri, and additional partners planted five hundred dunams of orchards; the attorney Otman Bushnaq (a descendent of the Bosnians who had reached Palestine) did the same. The attorney, writer, and national movement member A'adel Zu'aiter planted 120 dunams and the agronomist Jamal Hamad, also of a notable family in Nablus, planted one hundred dunams.[15]

Zarnuqa

A survey packed with details was conducted in 1934 by Moshe Smilansky's nephew, the statistician Ze'ev Smilansky who was also a member of the Ha-Po'el ha-Tsa'ir party. Ze'ev Smilansky had come to Palestine with the rest of his family in 1891 and lived with them in Hadera. After several years as the municipal statistician of Tel Aviv, he moved to Rehovot in 1927 when the municipality terminated his services. During the Mandate era, he placed many articles in the Hebrew press and published extensively in the journal that his uncle edited, *Bustenai*. Most of his articles concerned economic statistics and are of great value.

In 1934, writing under the nom de plume Z. S., he published a two-part article in *Bustenai* about the contribution of Jewish settlement to the economic and cultural "advancement" of the Arab *fallah*.[16] This article is unique in that it focuses on one locality near Rehovot and examines it from various perspectives. The locality at issue was the aforementioned Arab village of Zarnuqa. Smilansky's article stresses the economic and technological development that Zarnuqa was privileged to have undergone due to its proximity to the southern moshava. Adopting a slightly ironic tone in his reference to the development of agriculture in Zarnuqa, Z. S. praised the villagers in terms culled from the world of Ha-Po'el ha-Tsa'ir. Thus, the *fallahin* worked their land with "self-labor" and had "conquered" various occupations such as mechanized ploughing. Overall, Smilansky was amazed by the technological and economic progress that the village had made.

Z. S., like Lubman Haviv, found that the fulcrum of Zarnuqa's modernization was the sale of land to the Jews. Admittedly, Smilansky showed that villagers had been making a living by working in the moshava since the Ottoman era and "only by virtue of the income that they received from their Jewish neighbors could they save themselves from lives of poverty and deprivation." However, the upturn in Jewish immigration after the British conquest caused demand for land to increase, and land prices rose commensurably. Thus, not only effendis but also *fellahim* who owned smaller parcels began to sell land. "Those who lived near the Jewish settlement became aware that instead of the paltry income provided by the extensive farming that had been practiced since time immemorial, it was best to sell some of the land to the Jews and use the receipts to go over to irrigated crops, which economize on land and provide their owners with a better income." Smilansky stressed the improvement in overland transport between Rehovot and the city, the changeover to motor vehicles instead of camels or foot, and the improvement in the living conditions, nutrition, and clothing of the people of Zarnuqa. Even bicycles had become common means of transport in the Arab villages; "Dozens of workers come to [Rehovot] on bicycles [every day]," Smilansky exclaimed. Repeating a well-known Zionist argument, Smilansky showed that the population of Zarnuqa had grown by 102 percent between 1922 and 1934 while that of Hebron and Nablus had grown by only 8 percent and that of Gaza had actually contracted. This, he said, was due to the sources of livelihood that were available in areas bordering the Jewish settlements.

Most of Zarnuqa's economic growth traced to citriculture. Sales of land to Jews had injected capital into the village and created a steady increase in orchard area. In 1929, Zarnuqa had 1,122 dunams of citrus in thirty-two orchards. In 1934, following the steep increase in land prices—by more than 1000 percent relative to the decade-earlier level—the farmers of Zarnuqa had more capital in hand and invested it mainly in citriculture. Smilansky emphasized the change that occurred in orchard size. In the first years of planting, most orchards were spacious ones that were owned by the largest investors. By 1934, however, more than 60 percent of orchard owners had small holdings of three to twenty dunams apiece. The Arab growers had a comparative advantage over the Jews that lowered the cost of planting, Smilansky noted: About two-fifths of the owners continued to work as laborers in the moshavot until their orchards began to produce. Some even hired "cheap workers from the Negev" to work their orchards while they earned higher wages in Rehovot. They did not need to buy land; they earned extra income by growing vegetables between the rows of the young orchards; and "in most orchards, especially the small ones, self-labor is practiced and the women and children also regularly do various chores in the citrus orchards." The Arab growers also learned from their neighbors that they did not need to sink a well in every orchard; one could buy water from the owner of the orchard nearby.

Citriculture also created a source of living for people in more distant parts of the country, Smilansky reported. "In the citrus harvest season, convoys of dozens of camels laden with sacks of manure stream from the south of the country. This manure, worthless in the south, is a prized commodity here in the citrus area." Furthermore, Arab citriculture was developing no less than its Jewish counterpart; after all, nearly one thousand people had settled in Zarnuqa from faraway Gaza and Khan Yunes and were making a living by working in the orchards of the village. Concurrently, the ongoing Jewish immigration was exacerbating demand for working hands and was causing a labor shortage. Consequently, wages were rising not only in the Jews' orchards but in those of Zarnuqa as well.

Smilansky also emphasized the improvements in technology and knowledge that were reaching Zarnuqa in the wake of economic growth in the moshavot and the villagers' learning by example during their labor in the orchards of Rehovot (in what the research on technological diffusion calls the "demonstration effect"). Arabs were competing with Jews for jobs that had once been considered exclusively Jewish due to the expertise and knowledge that required. Arabs were mastering skilled orchard jobs, including mechanized labor ("working with the machines"), and quite a few Jewish growers employed Arab mechanics on a regular basis. Even the tractor work was no longer a Jewish monopoly, as Smilansky said it had once been. "The Arabs have begun to conquer this occupation, too!" Now Zarnuqa also had a

FIGURE 6.1. Citrus Marketing Board, 1940. From left to right: G. Sandford (Financial Secretary), Isaac Rokach, G. Walsh (Economic Advisor), and Said Beidas.

tractor, and its owner competed with the Jewish tractor owner for the land preparation and ploughing jobs that preceded planting of the orchard.

Another significant change occurred in the external appearance and technology of the Arab orchard, as though Aaronsohn and Soskin's article, written thirty years previously, made its way to Zarnuqa. Smilansky concludes:

> The Jewish farmers are good guides for our neighbors and do not charge tuition. Just a few years ago, one could find in the nearby villages wells with a beast turning the horizontal wheel. Now, all thirty-six wells in Zarnuqa are mechanized. Sarafand has orchards that receive electricity. Some orchard owners use the California irrigation method. Instead of planting the trees close together and in crooked rows, as in the previous method, our neighbors have begun to mark straight and widely spaced rows. The rule is, the progress of Jewish farming is leading to progress in our neighbors' economy: the development of the Jews' assets is leading to the development of our neighbors' assets.

PART IV

Growing Pains

SEVEN

Pursuit of Profit

THE PLANTING FRENZY AND ITS THREAT TO PROFITABILITY

THE UNDERLYING ISSUE in the expansion—and the contraction—of Palestine citriculture was its current and expected profitability. There are no confirmed and detailed data on the topic. One may, however, get a general picture from contemporary testimonies in the journals of the time, reports of experts who visited Palestine, internal documents that were not meant for publication, disputes among personalities and sectors in the Yishuv that burst into the open, and *en passant* statements made on various occasions. As stated above (chapter 1), the period of concern in this study may be divided into four main subperiods in terms of the profitability of citriculture:

1890–1919—the beginning of the industry, its initial expansion, and the crisis during World War I;
1920–1929—post-war rehabilitation and strong profitability;
1930–1934—reasonable profitability;
1935–1939—crisis and steep decline in profitability.

As stated in previous chapters, all the statistics indicate that citriculture was a very profitable industry during the Ottoman era. Its rapid expansion is perhaps the best evidence of this. According to our computation, profitability at the end of the Ottoman era stood at 24–58 percent of capital invested (see Table 4.3).

Between 1920 and 1925, as the industry was in its initial phases of recovery and reconsolidation after the devastation inflicted on it during World War I, new planting was relatively scanty. However, the profitability level began to rise in 1926. Journals of the time spoke of very strong profitability—

15–56 percent net profit to the grower—and the incentive to invest in citriculture increased tenfold. The Pardess reports, too, described the results of first postwar export season (1920/21) in "very glowing" terms. A year later, too, they stated that "the auspicious results of the 1921/22 season far exceed[ed]" those of the previous season. Two years later, in the 1923/24 season, the directors of Pardess wrote that the average price received had attained "the highest level in the history of the cooperative." The 1925/26 season, according to the Pardess reports for that year, was "one of the most successful that the orange industry has known since our cooperative was established."[1] Thus, it is no wonder that the press of the time, taking a cue from the biblical account of Joseph in Egypt, termed the seven years between 1921 and 1928 the "seven good years of Jewish citriculture."

The 1929/30 season marked the transition from strong profitability to reasonable profitability. The "on the tree" price of a case of oranges fell by 30 percent during that season (from seven shillings per case in previous seasons to five shillings). The profitability decline was worse for dealers than for growers, since the "on the tree" method gave growers a pre-assured price and subjected dealers to greater risk. This season marked the beginning of the end of the "seven good years," since the factors that would push citriculture profitability down in succeeding years were already evident by then, as the industry leaders knew. A sudden bountiful harvest surpassed the preseason forecast by about a million cases (40 percent more). Sorting and packing were poorly done, causing a high rate of fruit rot upon arrival at the port of destination. Spain turned out a large harvest. Finally, the great economic crisis that struck the Western world at the time vitiated purchasing power in the target markets. It is true that growers again received seven shillings per case of oranges "on the tree" in several shipping seasons after 1929/30, but in other seasons they received only the 1929/30 level or slightly less.[2]

Importantly, however, citriculture was still considered a safe and sound investment at this time, especially since the high prices in previous years had "cemented the image [of citriculture] as a very successful and profitable industry."[3] Thus, it comes as no surprise that during the reasonable profitability years, which, as stated, coincided with strong economic growth in the Jewish sector at large—a surge that was fueled mainly by impressive Jewish immigration and a massive inflow of private capital—the incentive for private investment in citriculture was strong. Citriculture was still thought to assure a higher rate of return than alternative investments. Indeed, the "planting frenzy" in Palestine citriculture largely overlapped the reasonable-profitability years.

Two years after the watershed 1929/30 season, a gradual but constant decrease in prices ensued: from between six and seven shillings per case "on the tree" to five shillings if not slightly lower. This happened for three main reasons: (1) an increase in supply, as the new plantations began to yield fruit;

(2) a decline in purchasing power in the British market due to the global economic crisis; and (3) the imposition by Britain of protective tariffs against citrus fruit from countries not included in the political definition of the British Empire, such as Palestine.

When the crisis erupted, as it did in the 1934/35 season, it embraced both national sectors of the industry. It was abetted by two main long-term factors: (1) the growing glut of citrus from Palestine and competing countries, especially Spain, and (2) the European countries' policy of "economic nationalism," which erected barriers against international trade in order to fight domestic unemployment. The potential crisis in Palestine citriculture became real in the 1934/35 season, when the German market for Palestine citrus contracted perceptibly. In 1933/34, Palestine exported to Germany 1.24 million cases of oranges, 24.4 percent of all Palestine orange exports that season. At the time, the German market was second only to Britain in its ability to absorb Palestine citrus. In 1934/35, in contrast, the German market received only 493,000 cases of Palestine citrus, 8 percent of the country's exports. Between the two seasons, the quantity of citrus produced increased by 25 percent. It was Britain, the largest (and the traditional) market for Palestine citrus, that absorbed this immense oversupply. Furthermore, the German market and additional markets in Central Europe, of which we speak below (even though their capacity for Palestine citrus was small in any case), suddenly closed their gates to the produce of Palestine. Thus, the glut problems steadily worsened and, concomitantly, profitability continued to slump.

In 1934/35, growers could still earn about P£5 per dunam of orchard, Moshe Smilansky claimed. By all accounts, however, 1936/37 was the worst season that Palestine citriculture had known to that time. The growers produced a record harvest (exports of 10.8 million cases, 83 percent more than in the preceding season and 47 percent more than the previous record, set in 1934/35). An especially rainy winter, damage to the fruit, and poor packing, however, resulted in a very high incidence of rot. The poor packing traced to the growers' wish to ship as much fruit as possible to market, irrespective of its quality. A severe upturn in competition between the Jewish and the Arab sectors at the time impaired profits even more. The sharp decline in earnings was abetted by two additional factors. The British government, in compliance with the terms of the Mandate, maintained an "open-door" policy vis-à-vis other member countries of the League of Nations (i.e., free trade with Palestine) and excluded Palestine from the imperial system of preferences. Exports from Spain flooded the market that season due to the civil war (see below).

In the 1936/37 season, according to the Jewish Agency Economic Research Institute, the orange market generated no net profit that could cover investments and offset a known extent of amortization. The grapefruit harvest began to become unprofitable during those years, mainly due to a large oversupply. Thus, it was alleged that "[c]itriculture at large derived no net profit from the grapefruit harvest and may have lost about half a shilling per case." Toward the end of the period, according to experts, 20 percent of growers had debts that exceeded the value of their orchards; 55 percent of growers had debts that did not exceed one-third of orchard value, and only 25 percent of growers had no debts or small debts only.[4]

A cloud of despondency began to descend over Jewish citriculture. The sense of crisis was aggravated by the awareness that the harvests would continue to grow with each passing year and that the problems of market closure and oversupply would only worsen. Although prices improved slightly in the next two seasons, they allowed the growers a small profit only and did not suffice to extricate the industry from its crisis.[5] Not until the summer of 1939, however, did the Jewish growers manage to formulate a common action plan to cut back on marketing and other costs. The beginning of World War II on the eve of the 1939/40 season disrupted all unification plans and thrust the industry into a totally new situation. Palestine went over to a war economy at the end of that season, and from December 1940 on it was the government that regulated the industry's affairs.

THE PROFITABILITY ISSUE AND ITS SPINOFFS

As I have shown, the high and medium profitability levels of Jewish citriculture (and of Palestine citriculture at large) during most of the period of concern in this study, coupled with the solidification of citriculture's image as a profitable industry, caused planted area to expand with great vigor during the Mandate era. Thus, it was inevitable that the industry would be threatened by, and would ultimately face in reality, a decline in profitability. The expected profitability slump could be countered in three principal ways: (1) cutting costs and increasing output per dunam; (2) expanding the markets; and (3) reducing competition and controlling the production end, that is, the supply side. Below we review what was done in these respects and ask how these actions helped to sustain the industry's economic strength.

Cutting Costs and Increasing Output per Dunam

In our foregoing discussion of this aspect of the growers' business activity, we summed up the data gathered thus far. Here we add a condensed table of current cost per export case from the orchard to the overseas market at two points in time: 1926, at the onset of the "planting frenzy," and 1938, at the height of the crisis; we shall add to that a profit and loss estimate per dunam of a bearing fruit orange orchard at these two points in time. We thus hope to present a coherent factual framework for continued discussion of citriculture marketing and profitability—which was, as stated, an important issue during most of the Mandate era.

As noted, the attempt to improve profitability has two components: increasing output and starting production at an earlier time, on the one hand, and trying to cut costs, on the other. In respect of the first component, I have shown that the Jewish growers did quite well, managing to shorten the time of receiving first returns on their investment and to increase their output. In regard to the second component, however, they had limited success only. Cultivation costs declined, but other costs—land, miscellaneous capital goods (mainly chemical fertilizer and packing materials), and interest rates—increased. Thus, the investment in the orchard became larger, annual cultivation became more expensive, and the overall FOB cost per export case climbed by 8 percent between 1926 and 1938. (See Table 5.5.) Table 7.1 below shows the costs of an export case upon arrival in the British market at two points in time, 1926 and 1938.

Table 7.1 aptly illustrates the increase in costs of harvesting, picking, and packing (mainly due to the increase in labor wage and an upturn in Jewish labor toward the end of the period at issue), the increase in prices of packing

TABLE 7.1
Current Costs per Crate of Oranges in Jewish Grove—
from the Grove to the Foreign Market (in Palestine Mills; 1926, 1938)

	1926		*1938*		*The Change*
Component	*Palestine Mills*	*1936 prices*	*Palestine Mills*	*1936 prices*	*1936 prices %*
Cultivation	114.8	86.8	85.5	84.2	(–) 2.9
Picking and packing	56.4	42.6	46.0	45.2	(+) 6.1
Materials	69.9	52.8	64.7	63.7	(+) 20.6
Haulage to the port	41.8	31.6	32.7	32.2	(+) 1.8
Taxes, et al.	39.5	29.9	24.5	24.1	(–) 19.3
Total to the ship	322.4	243.7	253.4	249.4	(+) 2.3
Sea transport and selling	215.2	162.7	156.5	154.0	(–) 5.3
Customs	—	—	45.0	44.3	
4% interest on foregoing	21.5	16.2	18.2	17.9	
Capital recovery, 8% for 30 years	108.1	81.7	104.4	102.8	
Total costs	667.2	504.3	577.5	568.4	(+) 12.7

Sources:

For 1926—Zagorodsky, *Citrus Book*, B, 227 (Hebrew). Zagorodsky's figures were converted to Palestine lira according to the rate of 1.02564 Palestine lira/Egyptian lira.

For 1938—Harry Viteles, "Survey of the Citrus Industry in Palestine," *Hadar* 11:1, 2–3 (1938): 22–23.

materials, and the upturn, albeit small, in the cost of overland haulage (mainly due to the changeover to rail transport and the extension of freight lines to Haifa). In contrast, day-to-day cultivation expenses (the second largest item in per-case cost, following marine transport) decreased perceptibly and taxes declined due to the aforementioned lowering of property tax. Interestingly, despite the increase in FOB cost per case, the cost of marine transport fell by 5 percent between 1926 and 1938, evidently because of the tremendous quantity of the harvest. Since the cost of marine transport to and sales in Britain (where most of the Palestine crop was marketed at the beginning of the period) was 38 percent and 33 percent of the total cost of an export case in 1926 and 1938, respectively (more than any other component of the cost of a case), the decline in this cost was able slightly to reduce the total cost of the case (if we disregard the cost of capital recovery for the moment). However, the tariff imposed on Palestine citrus in 1932 raised the cost of the case upon arrival in the British market by 10 percent relative to the 1926 level.[6]

This was in addition to the capital-recovery cost, which was higher per export case in 1938 than in 1926 because the orchards had become "younger" (due to the large extent of new plantations) and, therefore, less productive in

TABLE 7.2
Profit and Loss per Dunam of a Jewish-Owned Orange Grove, 1926, 1938

	1926		*1938*		*The Change*
Component	*Palestine Mills*	*1936 prices*	*Palestine Mills*	*1936 prices*	*1936 prices %*
Total costs per crate	443.4	335.1	367.9	362.1	(+) 8.0
FOB per crate	600.0	453.5	310.6	305.7	(–) 32.6
Net profit per crate	156.6	118.4	(–) 57.3	(–) 56.4	(–) 47.6
Net profit per dunam (Palestine lira)	19.6	14.8	(–) 5.73	(–) 5.64	(–) 138.1
Rate of return on investment	(+) 13%		(–) 5%		

Sources: Table 5.5.

For 1926—according to 125 export crates per dunam. FOB price according to Harry Viteles, "Expansion of the Orange Industry in Palestine," *Bulletin of the Palestine Economic Society* 3:1 (1928): 47.

For 1938—according to 100 export crates per dunam. FOB price according to A. Poshter, *The Development of the Citrus Marketing Cooperative Societies in the 1937/38 Season* (The Anglo-Palestine Bank, March 1939), 36.

the intervening years. The end result, in any event, was a 13 percent increase in cost per export case upon arrival in the British market. Concurrently, the return on a case of Palestine citrus decreased perceptibly—a development that everyone had expected but that few factored into their thinking. Indeed, the "bottom line," as Moshe Smilansky titled his article that pleaded for help, was a loss for the growers.

Notably, the 13 percent return on investment in 1926 reflects "conservative" figures. Other testimonies from the time indicate that an orchard produced more than 125 cases per dunam and that profit surpassed the level noted above. Be this as it may, we now examine the actions that growers took to prevent, solve, or at least alleviate the crisis that the table reflects so clearly.

Expanding the Markets

Market Players

The increase in supply entailed a search for new markets and further development of old ones. Three main factors influenced the marketability of Palestine citrus during this interwar period. The first and most important was a general increase in consumption of fruit, especially citrus fruit, in the West. The second factor was the uniqueness of Palestine citrus. The third was the competition that the Palestine fruit faced in the markets.

Western Europe (including Germany) was the main market for the citrus of Palestine. According to a study by Shulamit Mashke of the Jewish Agency Economic Research Institute, published in 1937, "the composition of nutrition of advanced European peoples" had been changing during the twentieth century; the emphasis was now on more "protective, vitamin-rich foods . . . at the expense of calorie-rich and vitamin-poor cereals." What is more, the increase in consumption of citrus fruits outpaced that of consumption of other fresh fruit after World War I. From 1933 on, however, citrus consumption declined slightly and the consumption of alternative fruit, especially bananas and apples, increased. It is obvious that the general uptrend in consumption of fresh fruit, including citrus, in the interwar years, was the largest contributing factor in the marketability of the growing supply of citrus. The industry leaders, however, were aware of the competing alternatives and knew that the market's ability to absorb the citrus of Palestine at a reasonable price was limited. Indeed, studies after World War II showed that demand for citrus in Britain (the principal market for Palestine citrus) was inelastic. In other words, an increase in supply had a stronger downward effect on prices than it had an upward effect on consumption, resulting in a decline in the total revenue.

Mashke also noted that the consumption of fresh food increased in tandem with income and commensurate with people's transition from blue-collar to white-collar occupations. She summed up her findings by noting that these figures demonstrated "the vast opportunities that exist for increasing consumption of fresh fruit by means of advertising and explaining needs without an increase in purchasing power."[7]

The second factor that affected the marketability of Palestine citrus abroad was the quality of the main fruit, the Shamouti orange. This issue is discussed above and needs no further attention here.

During the period at issue, Spain was Palestine's main rival in the citrus market in the winter season. Italy and the United States also shipped citrus to market during the winter and South Africa and Brazil did so near the end of the period, but the rival that counted was Spain. There were several reasons for this. First, Spain was the world's largest citrus exporter. In the 1926/27 season, for example, Spain exported 38.7 million cases as against only 3.5 million cases by Italy and 2.7 million by Palestine. The United States, although the world's leading citrus producer, exported only 500,000 crates to European markets that season. Even toward the end of the period examined here, Spain maintained its primacy among citrus exporters.

A second reason for the marginality of competition from other countries relative to Spain was that these countries, except for Italy, marketed most of their citrus produce in the summer. As the period drew to a close, Palestine, Brazil, and South Africa did step up their output and extend their shipping seasons. Therefore, the last portion of the export season in Palestine began to overlap the beginning of these countries' summer exports and the extension of the citrus season in Brazil and South Africa bumped into the beginning of the citrus export season in Palestine. These effects, however, were inconsequential in terms of the quantities that reached the market. Notably, too, it was actually the (cheap) Spanish orange and not the (expensive) "Jaffa" orange that gave the Italian orange (which commanded a price in between the two) its main competition. This further downsized the market segment for which Palestine and Italy competed. Moreover, for reasons of shipping cost and bilateral trade agreements, Italy exported most of its citrus to Central Europe and the countries along the Danube; hardly any went to Western Europe, the main market for the citrus of Palestine. Finally, the quality of the Italian orange declined steadily during the period at issue due to faulty cultivation and poor shipping conditions; this factor also mitigated the competitiveness of Italian citrus relative to that of Palestine.[8]

Thus, Spain remained Palestine's main rival in the markets of Britain and continental Europe. At this of all levels, however, an interesting development took place in the interwar period. The share of Spain in total global citrus exports went into decline, making Palestine in 1938 the second largest exporter. What is more, Palestine overtook Spain as the main supplier of citrus in several important markets, foremost the British.

The decline of Spain as a citrus exporter occurred primarily due to a crisis in Spanish citriculture that began in 1931 for reasons rooted in domestic and external factors.[9] Notable among the latter was the global economic depression and its outcome, a decline in the purchasing power of lower social classes—the main target population for the inexpensive Spanish oranges—and the raising of cross-border tariff barriers. The main internal factors that affected Spanish citriculture resembled some of the factors in Palestine: disorganization, inefficient sorting and packing, scanty government protection and inspection, and large-scale planting driven by excessive optimism after high-profitability seasons in the first two decades of the century. Additionally, Spain suffered frequent cold spells, a form of natural disaster that seldom struck Palestine. The frost damaged the trees and, in particular, the fruit, leaving the fruit misshapen and susceptible to fungus attack that eventually caused it to rot.

Most Spanish citrus is grown in the eastern part of the country, along the Mediterranean, including the vicinity of Valencia, which provided 80 percent of the Spanish citrus output during those years, and the areas of Murcia and Andalucia in the south. The predominant species of citrus was the orange, but it was grown in a very large number of varieties—up to thirty. The main advantage of Spanish citriculture over other countries' citrus industries was the lavish supply and very low production costs that allowed the Spanish orange to command a very low price. Spanish citriculture, like that in Palestine, was a bastion of private initiative. According to reports from the time, Spanish growers during the boom years had extremely high rates of return on their investments. Thus, it is no wonder that growers had a strong incentive to increase their planting. Indeed, the citrus area in Spain doubled between 1922 and 1939.

However, the combination of increase in supply—augmented by a mistaken choice of fertilizer that impaired the quality of the fruit—poor packing and sorting, and frequent cold snaps during the 1930s hurt the reputation and profitability of Spanish citriculture even before the Spanish Civil War broke out on July 17, 1936. Until January 1939, when Barcelona fell to Franco (two months before the surrender of Madrid brought the war to an end), Valencia was held by the Republicans. Spanish citrus exports did not cease during the civil war years, although they decreased and were plagued by irregularity. At the beginning of the 1936/37 season, the first season during the war, paltry quantities of oranges were shipped. At the middle of the season, the Republican government's need for foreign currency led to dumping of citrus, especially in the British market. The outcome—an unexpected bout of price cutting—contributed to the crisis of the 1936/37 season in Palestine citriculture. When Spanish exports contracted again in the 1937/38 season, the price of Spanish oranges in the British market rose. This fact did not escape the notice of the industry leaders in Palestine, who in various forums urged a corresponding (although voluntary) cutback in exports from Palestine.

THE MARKETS

Table 7.3 lists the main markets for Palestine citrus.

TABLE 7.3
The Main Markets and Their Share in Citrus Exports from Palestine (in percentages) (selected seasons)

Season	*Britain*	*Egypt*	*Germany*	*Rest of Europe*	*Others*
1925/6	88.4	8.4	No data	No data	3.2
1931/2	70.6	0.1	18.0	10.6	0.7
1933/4	62.4	—	24.4	12.3	0.9
1934/5	70.5	—	7.8	21.0	0.7
1937/8	63.0	—	1.6	34.9	0.5

Sources: William N. Hazen, *The Citrus Industry of Palestine* (Washington, DC: United States Department of Agriculture, Bureau of Agricultural Economics, December 1938), 2, 54–55; Knowles A. Ryerson, "The Horticultural Possibilities as Especially Related to Agricultural Colonization," *Reports of the Experts: Submitted to the Joint Palestine Survey Commission* (Boston: Daniels Printing Co., 1928), 298.

The data for the 1937/8 season are incomplete but representative; for the 1925/6 season the category "Others" includes Germany.

THE BRITISH MARKET

The share of the dominant British market in Palestine citrus declined gradually even though the supply of citrus that this market imported from Palestine grew considerably. For example, exports of oranges to Britain rose from 1.34 million cases in the 1925/26 season to 6.61 million cases in 1936/37—nearly a fivefold increase within ten years.[10] Notably, too, Palestine citrus made inroads in new markets in Britain itself. Until 1914, Jaffa oranges were marketed almost exclusively to Liverpool, whereas by the eve of World War II they were making their way to Glasgow, Manchester, and, especially, London. London was considered the bastion of the Spanish orange; the attempts to claim a market share there at the beginning of the period were hesitant and unsuccessful. However, as harvests increased, there was no choice but to invest in a breakthrough there. Shipments to London increased from 1931 on, and by the late 1930s the London market accounted for one-third of Palestine oranges exported to Britain. Another third went to Liverpool during those years and the share of Manchester declined from 16 percent in the early 1920s to 4 percent ten years later. Notably, too, the London market was more important than the mere quantity of citrus that it received. This market evinced brisk demand for large oranges, the sort typically produced by young orchards. Thus, London was the market of entry for the steadily increasing harvests of the new orchards.

As stated, one of the main trends in the development of the British market in the interwar years was the decline of Spain as the chief supplier of citrus and the ascendancy of Palestine to that status. In 1930, Spain provided 71 percent of citrus in the British market and Palestine furnished 13 percent; in 1937, the respective shares were 30 percent and 36 percent. In the 1931/32 season, when the first yields of the new plantations of Palestine reached the market, this fruit began to trickle down to the lower British social classes. Concurrently, its price fell almost to the level of the Spanish fruit. Were it not for the "external factors"—the relative and absolute decline in Spanish exports to the British market and the damage to Spain's reputation—the crisis of citriculture in Palestine presumably would have erupted before the 1936/37 season.[11]

By the end of the period, as noted earlier, citrus fruit from South Africa and Brazil was giving that of Palestine a measure of competition. British imports of citrus from Brazil, in particular, increased impressively—from only 3.5 percent of oranges imported in 1930 to 14 percent in 1936. After South Africa and Brazil extended their citrus shipping seasons, exports of Palestine citrus were impaired, especially in the December holiday season, when prices were especially good. For this reason, although the Palestine fruit was not yet fully ripe in December, the shippers rushed their fruit to the British market even if it was green and sour. The Brazilian season, in contrast, ended in December. Thus, this fruit reached the markets at peak ripeness and was obviously preferred by customers. Even so, the competition was marginal. In December 1936, for example, Palestine shipped 1.12 million cases of oranges to the British market, Spain 790,000, Brazil 109,000, and South Africa only 11,000. If so, exports from Palestine outpaced those from Brazil by a ratio of ten to one. What is more, the beginning of the South African and Brazilian orange export season overlapped the end of the Palestine season (April), and those countries' exporters fell victim to the same "ruse" that they had tried to pull at the end of their season.

Another noteworthy aspect of the British market was the gradual change in the share of this market in the exports of the two sectors of Palestine citriculture. This happened because Palestine citrus commanded lower prices in the British market than in continental Europe. As stated, production costs were lower in Arab citriculture than in Jewish citriculture. For this reason—and because of its more entrepreneurial nature—the Jewish sector of the industry was the first to hunt for new markets that would provide a better return. By the end of the period (1938/39), as I explain in detail below, the Jewish sector shipped an estimated 55 to 60 percent of its produce to the British market while the Arabs sent 80 percent of their yield there. This trend was even more pronounced in the efforts of the Jewish citrus cooperatives, which were sending less than 55 percent of their produce to Britain by the end of the period.[12]

Other Markets

As mentioned above (Table 7.3) continental Europe was the second most important market for Palestine citrus, trailing only Britain. Palestine produce also reached markets outside of Europe. However, with the exception of Egypt in the 1920s, export of oranges to non-European destinations—such as India, Australia, Ceylon, and Canada—failed to exceed even 1 percent of total Palestine citrus exports during the period at issue. For the most part, these exports made a contribution of the exotic kind. Exports to Europe, in contrast, increased in proportional (and of course, in absolute) terms throughout the interwar period, from a few percent at the beginning of the period to as much as 40 percent by its end. The fact that eighteen countries in continental Europe were important markets for the produce of Palestine gives an indication of the citrus producers' success in developing new markets for their burgeoning harvests in order to keep their industry profitable. If an existing market were expanded beyond a certain quantity of fruit, as occurred in Britain (especially after the 1934/35 season), prices might tumble. Opening a new market, in contrast, with a relatively small supply of produce relative to the old markets, and with a high-quality product such as the Shamouti, usually assured high prices. This is evidently the main reason that exports even to markets that were protected by barriers—quotas, high tariffs, currency restrictions, etc.—did not cease and exporters from Palestine found it worth their while to continue shipping to them. Generally speaking, as stated above, contemporary statistics show that the price level in Europe exceeded that in Britain. The decline in the centrality of the British market in Palestine citrus exports gives the best possible evidence of this fact.

One may divide the markets of continental Europe into two main groups: free-trade and restrained-trade. Trade barriers were typical of the time (especially in the Great Depression years), which was noted for the contraction of many markets due to the "economic nationalism" policy that many governments adopted. Since the contraction of markets coincided with the steady increase in supply of citrus from Palestine, the growing pains of Palestine citriculture worsened. Countries that allowed free trade in citrus became the principal markets for Palestine fruit. The reason was simple: The "open door" policy expressed in the terms of the British mandate for Palestine ruled out a policy of selective reciprocity in trade with countries on the continent. Notably, too, the proximity of Spain and Italy to the continental markets gave these countries a salient advantage in their trade relations with countries that restricted citrus imports. Thus, Spain claimed a lion's share in the markets of France and Germany, and Italy had a large advantage in its relations with the countries of the Danube region and, after 1935, with Germany as well.[13]

The increase in exports to countries that applied restrictions, was due mainly to the growth of the Polish market, where the tariff was lowered slightly in 1935. The Dutch market developed mainly after the industry lost its German market in the 1934/35 season. The Dutch also bought much of the young (large) fruit, which was less in demand than small fruit in most markets.

Summing up, the market expansion policy proved itself as long as the conditions for free transnational trade did not worsen. However, when tariffs were applied in the main market for Palestine citrus, Britain, and when the German market was almost totally closed off later, a glut crisis became inevitable and the leaders of the industry—and a fortiori the rank-and-file growers—were forced to seek other ways of extricating themselves from it.

EIGHT

The British Mandate Government's Policy toward the Citrus Industry

REDUCING COMPETITION AMONG growers and controlling the quantity of fruit exported were considered tried-and-true ways of stanching the steep decrease in profitability during the 1930s. Indeed, most players in the industry—including the leading executives of the banks, which provided most growers with the credit they needed—evidently found acceptable these two "remedies" for the ailments of Jewish citriculture and of Palestine citriculture at large. During most of the period of concern in this study, however, support for restraint of competition and supply was mostly theoretical; when it came to implementation, many difficulties arose. There seem to have been two entities that could undertake the daunting task of reducing supply and competition: the government and the growers themselves.

Ostensibly, the Mandatory Government could intervene in citriculture at an early phase and avert the crisis that beset it in the late 1930s. Indeed, proposals to this effect, such as the imposed unification (cartelization) of all citriculture in Palestine (Jewish and Arab alike), had been bruited since the late 1920s. However, the government's policy toward the citrus industry was characterized by the same ambivalence that was typical of its economic policies vis-à-vis other industries. On the one hand, the government operated within the legal framework of its Mandate and applied a liberal-capitalist worldview. On the other hand, the realities of Palestine, which the government and its officials faced when they set forth to implement this policy, induced them to overstep the broader legal and ideological framework in several fields.[1]

Furthermore, the citrus industry, like the economy of Palestine at large, was subordinate to the legal, administrative and economic framework that the British regime had set forth and was profoundly influenced by it. It is within this ambit that we shall examine the implementation of government policy toward the citrus industry of Palestine.

Under the basic conditions that the government of Palestine had created for economic activity and, in particular, for citriculture, two main elements deserve emphasis. First, as noted above, citriculture operated under the constraints of the "open door" policy that was set forth in Article 18 of the Palestine Mandate. It obligated the mandatory government to place no restrictions on trade between Palestine and any member state of the League of Nations

The leaders of Palestine citriculture seem to have wasted no opportunity to petition the authorities against the damage that this policy cost their industry. They stressed above all the inability to force major importers to Palestine that in their respective countries put restrictions on citrus imports from Palestine, to give the Palestinian citrus industry some preference. They even suggested various remedies. However, the government did not change its policy until the beginning of World War II. The government claimed that amending Article 18 would reopen all the articles of the Mandate to rediscussion; this was certainly a main consideration in its unresponsiveness. Furthermore, until the crisis season of 1936/37, the government considered the citrus industry a very profitable one and saw no need to intervene in it or protect it in any way.[2]

The second important element was the overall development of Palestine, especially in transport and communications infrastructure. The director of the Mandatory Agriculture Department, summing up a decade of interrelations between the citrus industry and the government between 1928 and 1938, began his remarks by noting the government's contribution to the development of infrastructure. Especially noteworthy in this context were the expansion of the rail system, which, by 1929, connected all citrus centers with Haifa; the expansion of the road system (including auxiliary roads that linked citrus areas with main roads); and the construction of Haifa port. Obviously, too, the development of communications—at the domestic level and between Palestine and abroad—also did much to further commercial activity in citriculture.[3]

The second level of government involvement in citriculture was bureaucratic and manifested foremost in relations between the Agriculture Department and the industry. The department interacted with the industry by means of a special advisory committee, composed of British, Arab, and Jewish members, that operated more or less uninterruptedly during the period at issue. This committee, titled as the Citrus Fruit Committee, was part of a matrix of advisory committees that the Agricultural Department established in various matters. These committees reported to the General Agricultural Council, established in February 1931, and the council, in turn, presented the government with various recommendations for action in its field of competence. The problem was that the council's powers were limited ab initio to the coordination of agricultural research and education. The powers of the Citrus Fruit Committee, too, were advisory only. Here was another manifestation of the government's policy of minimal intervention.

Just the same, the Citrus Fruit Committee played several important roles in citriculture activity. First, it provided the three main players in citriculture—the government, the Arab industry leaders, and the Jewish industry leaders—with an official and permanent place to meet. Another important function of the committee was to discuss the main issues of concern to the industry. They included inspection of citrus exports, war on pests and diseases, occasional citrus trade fairs that were meant to provide growers with information, citrus advertising in foreign market, citrus taxation, coordinating of fruit shipments, and the establishment of subcommittees for specific matters such as transport and haulage within orchards and thence to export markets. Thus, the committee served as a setting for the mitigation of disagreements among the three industry players and, above all, for dialogue between the government and the growers.

The government helped the citrus industry in various additional respects. It invited experts from abroad to examine general and specific aspects of the citrus industry; established four experimental stations, albeit small ones, which conducted limited experiments in the care of citrus and created a permanent exhibit of appropriate care of fruit; subsidized the Jewish Agency's research station on the condition that the station and the results of its experiments be made available to the Arab growers as well, co-funded pesticide campaigns; combated an invasion of locusts in 1930; and organized several special citrus exhibitions.[4]

Concurrent with the administrative activity and usually in coordination with, and with the knowledge of the Citrus Fruit Committee, the government enacted several regulations and laws that controlled certain areas of citrus industry activity. For example, the importation of rootstocks and scions without careful quarantining was prohibited to thwart the introduction of pests and diseases. In 1934, compulsory licensing of nurseries was introduced to prevent the spread of pests and diseases and to assure the quality of rootstocks and scions. However, the government's two principal activities in citriculture in this context were the enactment of marketing and advertising regulations.

The Fruit Export Ordinance went into effect on January 1, 1928.[5] Until then, the quality, size, shipping schedules, and other aspects of exported fruit were totally unregulated. In 1926, the Imperial Economic Committee on Fruit called the attention of Lord Herbert Onslow Plumer, the High Commissioner, to the fact that the good reputation of Jaffa oranges in U.K. markets was at some risk due to the decline of the quality of some recent shipments. In response, Lord Plumer appointed the Fruit Commission, in which all sectors of the industry were represented, to tender recommendations for improvements in the preparation and marketing of the citrus fruit of Palestine. The commission's principal recommendation was to control the quality of fruit exported from Palestine by establishing an inspection service. Since the recommendations initially aroused stiff opposition—mainly among Arab dealers—the first ordinance imposed minimal requirements: stating where inspection was to take place and making a general stipulation about fruit quality. During that season, 1927/28, it was impossible to reject a shipment without the exporter's consent and to force an exporter to repack a faultily packed shipment. The very next year, however, the regulations were toughened somewhat and a beginning-of-season date was set (November 1) before which exports were prohibited. The inspection regulations became increasingly explicit and stringent over the years: A size for export cases was stipulated and special regulations for grapefruit were passed. From 1932 on, exporters had to register themselves and their trademarks, and the number of trademarks per exporter was limited. For the 1936/37 season, exports of large fruit from new orchards were restricted, and the enclosure of packing slips was made compulsory. After the poor 1936/37 season, in which the percent of fruit rot was especially high, the regulations were toughened further and government involvement in citrus exports became more intrusive. From now on, not only a shipment of fruit but also an orchard, a packing house, or a motor vehicle could be disqualified if its conditions of hygiene and disinfection were poor.

One may construe the escalation in government involvement by means of the Fruit Export Ordinance in two ways. First, it seems clear that the domestic consideration among the bureaucrats (a local patriotism of sorts) had gathered strength. Second, this involvement did not necessarily contravene the overall policy of the Mandatory Government. This is because the government considered it a goal of the highest order—due to its role as custodian of Palestine and its inhabitants and for internal security and internal relations reasons—to go to lengths for the country's development and welfare. Thus, it regarded the citrus industry, as the most important export industry that was shared by both peoples that inhabited the country, as highly deserving of encouragement and assistance. For this reason, as the years passed and the profitability of citriculture declined, one might consider the government's involvement in determining the size, quality, and timing of fruit exported from the perspective of protecting the industry against losses or collapse. An additional manifestation of this attitude was provided after the 1936/37 season, when the rural property tax for orchard owners was lowered from 825 Palestine mils per dunam to only 500 mils. Furthermore, the customs policy was revised and in advance of 1938 the government undertook to refund to the growers customs that had been collected on imported wood, nails, and baling wire that were used in the manufacture of orange cases.

The Fruit Inspection Service generally drew uncomplimentary criticism. At Jaffa port, through which most export shipments passed, the inspection conditions were particularly poor. Everyone knew that it was very easy to outwit the few inspectors who worked there and to slip damaged fruit into the flow of millions of cases that passed through the port each season. Fruit that was shipped to market via Haifa, nearly all of which came from Jewish-owned orchards, passed stricter inspections—usually at the main railroad stations. Even so, Jewish industry leaders claimed continually that the Fruit Inspection Service was inadequate in scale and in the toughness of its measures. Indeed, most Jewish cooperatives did not trust the government inspection system and conducted stringent inspections of their own in shelters at the railroad stations. Their intention was clear: Unable to compete with the lower cost of a

FIGURE 8.1. At a citrus exhibition in London: The Pardess exhibit, 1930s.

case of Spanish oranges or of an export case from the Arab sector of Palestine, Jewish exporters attempted to boost turnover by limiting the quantity and improving the quality of their export fruit.

The advertising regulations went into effect in advance of the 1933/34 season, after delays and postponements that stemmed from disagreements between Arab and Jewish growers about criteria for collecting the tax that was to finance advertising in European and British markets.[6] The Arab growers favored a tax based on land area; the Jewish growers, whose orchard area surpassed that of the Arabs by then, preferred a tax per case. Ultimately, the government decided on a tax per export case. The importance of advertising as a way to expand the markets had been stressed since the beginning of the Mandate era. However, until harvests grew and industry profitability was jeopardized—and until the authorities took a more localized view about the way the country and its economy should be managed—regulations concerning advertising were not enacted. It is true that in the late 1920s the British government included the citrus fruit of Palestine in a general advertising campaign, under the slogan "Eat More Fruit," designed to increase fruit consumption. However, not until 1934, as stated, did a specific advertising campaign for Palestine oranges get under way—posters on outer panels of buses carrying the slogan, "Jaffa for Juice." Additional advertising took place in newspaper advertisements, radio broadcasts, and show windows of fruit shops. After the initial promotion season, the advertising budget grew (from P£ 6,700 in 1933/34 to P£ 40,000 in 1937/38) and the Palestine orange was advertised in diverse media in some twenty countries. Although the industry leaders disagreed about the effectiveness of the advertising, everyone agreed that advertising the Jaffa orange in the British and continental markets should continue and should intensify. Here, as in other aspects of Palestine citriculture during the Mandate era, there were early advocates of action to develop the advertising effort. Those voices, however, were not heeded in any substantive way as long as handsome profits were being made. The only difference now was the government's involvement in managing the advertising project.

Summing up, the government's involvement in citriculture was minor, and even though it gathered momentum during the period at issue—especially in inspection and advertising—it did not suffice to avert the crisis of the late 1930s. We now examine the ways in which the growers themselves attempted to limit citrus supply and internal competition and ask how well they succeeded in doing so.

NINE

Attempts to Establish a Cartel

THE CALIFORNIA MODEL

PROPOSALS TO ESTABLISH a growers' cartel preceded the onset of the great citriculture crisis in the mid-1930s. By the early twentieth century, various parties involved in citriculture (Jewish and German), especially bankers, touted the great advantages that the establishment from such a cartel would deliver. It was not just the special situation of citriculture in Palestine that touched off discussion about reorganizing the industry in the form of a cartel. The growers were well aware of quite a few cartel models that clearly demonstrated—in the opinion of those who favored cartelization—the virtues of the cartel method of marketing over the conventional method in Palestine at the time. The most conspicuous of these models was the California one.

Previous sections of this study discussed various aspects of the California model. Here we should stress that, despite the well-developed technology of California citriculture and the permanent relationship between growers there and the research departments of the University of California and of the government, the very core of citriculture organization in California was the producers' cartel, the California Fruit Growers Exchange. The path to the formation of the exchange was not short. Not until 1906, after several organizations of various types had risen and fallen, did the growers of California manage to establish the marketing entity that would assure them the best results. The CFGE was a statewide corporation that brought growers' associations at the local, regional, and state levels under one roof. It marketed 75 percent of California fruit in 1928 and 80 percent in 1937. It managed marketing for its members, sent representatives to the markets, and controlled the quantities of fruit to be sold and the areas to which the produce would be sent. The exchange allowed only a small number of brand names for each species of citrus, required the use of cooperative regional packinghouses, and enjoined members from packing or shipping fruit by themselves. Due to its cartel power, its members were able to obtain materials and services at greatly reduced prices and command higher prices for their fruit in the markets.

In 1912, as noted, Baron de Rothschild's agronomist and a member of the executive committee of the Pardess cooperative, Abraham Brill, was sent to California to study its citrus industry. Returning to Palestine, Brill lectured on what he had learned in California about locally unknown methods that would improve packing and marketing. The minutes of Pardess meetings in late 1913 show that Brill attempted to convince the members of the vast advantages that they would gain if they unified their industry along the CFGE lines. Expenses and wages were twice as high in California as in Jewish citriculture, Brill claimed, but "their prices are twice ours." The main reason was a central organization that handled the sorting, assured good and attractive packing, and ensured appropriate arrangements for delivery. The members of Pardess tried to cram the greatest possible number of cases into each shipment "without sorting the nice fruit" and used the local packing method, which placed several sizes of fruit in one case. In California, in contrast, growers sorted carefully and placed uniform fruit in size and genus in one case. Thus, the fruit that reached the market was attractive in packaging and form, and although its quality did not match that of the Palestine Shamouti, it commanded twice the price. Noting that the Jewish growers and Pardess had just come through two bad seasons, 1911/12 and 1912/13, Brill argued that without unification the industry would wither away.[1]

Thus, the CFGE model became an organizational ideal for the Jewish citriculture industry. As early as April 1913, after two seasons of gluts and losses, the Anglo-Palestine Bank (APB) and the Deutsche Palästina Bank sponsored the formation of a cartel with the German and the Arab marketers. The initiative got as far as the preparation of articles of incorporation. However, it did not turn out well—mainly, it seems, due to disagreements between Pardess and the other Jewish growers about the role of Pardess in the proposed unification. The beginning of World War I about a year later throttled any further initiative of this type.[2]

During the Mandate era, the question of unifying Jewish and Palestine citriculture was debated every now and then. In February 1928, for example, Professor Clark Powell of South Africa, an expert who visited Palestine on behalf of the Mandatory Department of Agriculture, recommended the California method as the surest vehicle that Palestine citriculture could use to attain stability while continuing to expand. His recommendation was not taken up. Two years later, following the poor 1929/30 season, the banker Harry (Zvi) Viteles published a detailed proposal for the organization of the "orange industry in Palestine" following the California model. It was not by chance, of course, that the directors of banks that had given the growers much credit were concerned about the decline in the industry's profitability. However, it was equally noncoincidental that Viteles and the bank that he headed, the Central Bank for Cooperative Institutions in Palestine, were the parties that brought up the proposal for unification following the California model. After all, one of their main goals in Palestine was to establish cooperatives that would help medium and small entrepreneurs sustain themselves while remaining independent and "eager."

Attached to Viteles's proposal, which was published in the Hebrew-English monthly journal *Hadar*, was a preface by E. R. Sawer, director of the Mandatory Department of Agriculture and Forests. Sawer viewed the decline in citriculture profitability soberly and traced it to factors that had nothing to do with external conditions. In his opinion, some decline in profitability was inevitable and would probably not be limited to the 1929/30 season. Presumably, he continued, it would recur in the seasons to come due to the burgeoning supply of produce and the lack of increase in demand. "Let us admit," the director of the Department of Agriculture wrote, "that there is no further hope for 'handsome prices' and net profits of 25 Palestine pounds or more per dunam; some reduction in prices is unavoidable." Sawer heartily endorsed Viteles's proposal, considering it an appropriate way to avert a crisis that might erupt within "three years or so." Organizing the industry as Viteles recommended, Sawer said, would improve the poor sorting and packing methods that were bringing down the price that the Shamouti orange would otherwise receive. The Viteles plan would also allow the industry to apportion and tailor its shipments appropriately and to acquire tools to develop and exploit new markets that were waiting "on our very doorstep, in Eastern and Central Europe."

It is noteworthy that even though one might think, in view of the preface by the head of the Agriculture Department, that Viteles's proposal dealt with both national sectors of the industry, Viteles spoke of organizing Jewish citriculture only. The director of the Central Bank for Cooperative Institutions made only vague and marginal reference to the Arab sector. Apparently both Viteles and Sawer were highly skeptical about the possibility of unifying the two wings of the industry. Viteles also thought it unlikely that the Jewish growers could come together. In his opinion, which he expressed vehemently as early as 1928, only government intervention could make the unification of the citrus industry a reality. However, the government stated explicitly during most of the period that the growers should attend to the matter themselves and, as I have shown, refrained permanently from intervening. Even Sawer, in his preface, expressed the same outlook, that is, theoretical encouragement but no commitment to real action.[3]

Various factors thwarted the unification of Palestine citriculture during most of the Mandate era. The most important of them, it would seem, were the industry's individualistic nature, its conspicuous predisposition to private enterprise, and the strong profitability that it maintained during most of the era. These factors were valid in both the Arab and Jewish sectors of the industry.

Indeed, a serious outside observer such as Knowles Ryerson, who, as we recall, visited Palestine to examine the domestic citrus industry as a member of the Zionist Organization delegation of experts, summarized his impression of the matter as follows:

> The history of the development of cooperative effort in other countries indicates that progress comes only on the heels of disaster. It is not thought that the case will be different in Palestine. Increased production and increased costs of production in Palestine, increased competition with other countries will inevitably bring about a reduction of net return. Present high returns have brought about a planting boom such as Florida and California have both experienced in the past. Such a period of rapid expansion and wasteful practices brings on its own readjustments, abrupt and disconcerting as they may be. It is during such times that the vital importance of cooperative effort overshadows lesser differences and progress can be made.[4]

In retrospect, Ryerson's keen observation about human nature and the capitalistic, individualistic nature of the industry was quite on target. Thus, unification came about only after the citrus industry was teetering on the brink of an abyss.

JEWISH CITRICULTURE MOVES TOWARD UNIFICATION

Mancur Olson's model of the Logic of Collective Action seems to describe adroitly the progress toward cooperation in Palestine's citriculture industry. Most growers acknowledged the need for a cooperative organization but lacked an economic interest strong enough to pay for the "public good" that the organization was to deliver. Each grower expected someone else to assume the cost of establishing the unified organization and of the immediate loss of earnings that this step was expected to create. Only when profitability began to decline did an organizational process ensue in which the price of entering and sustaining the organization was lower than the return flowing from participation in it.[5]

When we examine how the Jewish growers adjusted to the cooperative form of organization, we find several main tendencies: (1) a transition from an individual organizational pattern and marketing by means of dealers to organizational coalescence within a cooperative framework and cooperative marketing, although not yet under one countrywide umbrella; (2) pressure for unification on the part of several industry leaders before the conditions matured, resulting in the formation of several stillborn organizations; and (3) fragmentation among the cooperatives, in view of ideology (the Labor bloc vs. the rest of the Yishuv), proposed answers to the unification question, and economic conflict of interests, foremost between the "old citriculture" and the "new citriculture."

From Individual to Cooperative Marketing

Until the early 1930s, citrus fruit in Palestine was marketed in three principal ways: sale to dealers before the season ("on the tree"), marketing by dealers on a commission basis, and cooperative selling by means of Pardess.

The most prevalent method until the early 1930s, practiced in Palestine even before Pardess had been founded in 1903, was sale *"on the tree."* In the late 1920s, 70 percent of Palestine citrus was traded in this way. This trade was administered mainly by several dozen Arab dealers in Jaffa, augmented by a few Jewish dealers. A buyer of fruit "on the tree" visited the orchard whose fruit he wished to buy several months before the season began and estimated the yield. The total sale sum was determined on the basis of this estimate and the presumed future price in Britain, where most of the fruit acquired in this manner was marketed. Payment was usually made in three installments: one when the contract was signed, another at the beginning of the season, and the third in the middle or toward the end of the season. The purchase was financed from three main sources: the dealer's own capital, bank loans, and, in most cases, an advance from a broker at the Liverpool fruit exchange with whom the dealer had a relationship. It was usually the dealer who had the fruit picked and packed; the grower's responsibilities were limited to irrigating the orchard, tending it, and cultivating it until the harvest began. Sometimes the contract stipulated that some of the picking and packing work be done by Jewish labor.

The drawbacks of "on the tree" buying were evident to one and all and were often discussed in the contemporaneous literature. The Achilles' heel of the method was its speculative aspect, manifested in estimating the harvest and, most crucially, the expected price—all several months before the fruit ripened and the harvest season could begin. Furthermore, since the dealer's main interest was to obtain an immediate profit that season, he was not stringent about the quality of packing and the reputation of the fruit that he marketed. If these dealers engaged in additional fields of commerce, as they often did, they had even less of an interest in maintaining the long-term reputation of the Jaffa orange. Thus, the marketing of Palestine citrus suffered from all the drawbacks of nonstandard and unorganized marketing: emphasis on the quantity marketed and neglect of quality; failure to coordinate shipments, gluts in some markets and shortages in others; premature marketing of green fruit and, in mid-season, marketing of *berara* as high-quality fruit.

However, sale of fruit "on the tree" also had several significant advantages. The entrenchment of the method, which in the Arab sector remained common even at the end of the period—by which time the Jewish sector had almost totally abandoned it—proves how expedient it was for both dealers and growers. Importantly, despite the seemingly speculative nature of this method, in the long term the expertise of the broker and the dealer, who operated together in this line of business, seems to have eliminated much of the margin for speculative error. Furthermore, at the beginning of the period sale "on the tree" often led to higher net profits than other marketing methods. Even when this was not the case, the grower was still tempted to maintain his or her independence and not to take on the liabilities and mutual guarantees that came with membership in a cooperative society. The "on the tree" method allowed the grower to obtain convenient financing for cultivation during the lengthy months between seasons. Last but not least, psychologically, this method eliminated any doubts that might beset the grower during the tense time of the season by shifting the risk to the dealer and his financiers. As Isaac Rokach put it, with the "on the tree" sale method one could "swindle an Arab or a Jew, get capital from him, and sleep through a night of storm and hail."[6]

The second method practiced in Palestine, although on a very small scale until the middle of the 1930s, was *commission sale*. According to this method, fruit dealers paid the growers the full consideration that they received for their cases in overseas markets and subtracted their commission. Commission sale was usually practiced by agents of brokers who sat in Jaffa or companies and dealers who represented brokers. For example, Zalman Jacobson and Lazar (Elazar) Rabinovitz's company, Associated Orange Growers, Ltd., agents in Palestine for the brokerage firms Connolly Shaw, Ltd., and Russell Turnbull & Co., Ltd., which operated throughout England, Ireland, and Scotland, regularly engaged in "on the tree" purchase and commission selling. When the commission method was used, it was the grower who took care of picking, sorting, and packing, while the dealers handled shipping. They also inspected the quality of packing because, in contrast to the "on the tree" marketing method, in commission selling the broker had an interest in assuring the shipment of fruit of a known level of quality. The incentive for the grower was a higher estimated return per case of fruit than the "on the tree" method could provide, because the fruit was sorted and packed better and marketed by brokers of repute.

In October 1931, for example, Associated Orange Growers signed an agreement for the shipping of oranges on a commission basis. The brothers Ahmed and Ismail Shneir, dealers from Jaffa, undertook to ship by means of Associated Orange Growers and their brokers fifty thousand cases of oranges during the 1931/32 and 1932/33 seasons; to pay the company fifteen Palestine mils per case in commission for handling the fruit in Palestine and 2 shillings per case for sale in England, and to accept the company's inspection of the packing and sorting. Incidentally, the Shneir brothers—as the contract makes clear—marketed by means of Associated Orange Growers not only fruit from their own orchards but also fruit purchased "on the tree" by taking an interest-bearing loan from Associated Orange Growers.[7] This example illustrates the important role of the English broker as a major financier of the Palestine citrus trade. Apparently the brokers of Associated Orange Growers furnished the Shneir brothers with the principal of the loan, and it became possible to sign the entire accord on the basis of the transaction that would be undertaken with them during the season.

The third marketing vehicle that growers commonly used at the beginning at the period was the *Pardess cooperative*. Pardess was established in Petah Tikva in 1903 by a group of growers and dealers from the "Mother of Settlements" (as Petah Tikva was known), Jerusalem, and Jaffa, with the active participation of the ICA. This producers' cooperative was established following the agricultural cooperative model of Friedrich Raiffeisen of Germany (1818–1888). Raiffeisen, whose contemporary in the urban domain was Hermann Schulze-Delitzsch (1808–1883), established cooperative societies in rural areas of Germany that provided members with credit, marketed their farm produce, and provided raw materials and goods for the production process. The financial transactions of the Raiffeisen-type cooperatives were backed by unlimited guarantees put up by guarantors and the peasants' landholdings. The Raiffeisen cooperatives quickly went countrywide, and at this level, too, they engaged in credit, marketing, and supplies for members. By setting up cooperatives in rural and urban areas, Raiffeisen, Schulze-Delitzsch, and their successors learned how to establish financial institutions that provided small and medium entrepreneurs, rural and urban, with capital for their current operations. Furthermore, their use of the scale economies of cooperative enterprise reduced marketing and producing costs.

Notably, the Yishuv was aware of the producer-cooperative models—Raiffeisen's and others—at the time Pardess was established. Not only were Jews of European origin (from both the eastern and the western parts of the continent) familiar with the cooperative movement; many of them, or their relatives and associates, had taken part in it in their places of origin.

Pardess was established mainly to improve the Jewish growers' return that the Jaffa fruit dealers gave them for their produce. "Those in the know," the minutes of the founding meeting stated, predict "a great, glowing future [for the orange industry], especially if we find a way to deal in the commercial aspect by ourselves, too, [an aspect] that depends on the sale of our fruit and that is currently controlled exclusively by the non-Jewish dealers of Jaffa." Accordingly:

> We have resolved unanimously to establish a society, since we, too, have learned from experience that a collective is better than individuals. We

> have undertaken to act on behalf of all members of the cooperative to the extent of each person's ability, and it is our hope that our unity will enable us to act, succeed, and thrive.[8]

In 1902/03, the season when Pardess was founded and made its first shipment, the cooperative had twenty-eight members. There seems to have been an expectation of greater profits through marketing by means of the cooperative than by means of the Jaffa dealers; this was an important incentive for growers to join. By 1909/10, growers were joining Pardess at a rising pace and in a meeting of the cooperative on June 29, 1910, the leader of Pardess since its founding day, Shimon Rokach, listed the advantages of associating under the Pardess umbrella: Everyone related to the industry—dealers, shipping agents, banks, vendors—respected Pardess and gave it service under preferential terms. "In Jaffa" Rokach explained, "some dealers have many more cases than our society does; nevertheless, they are not as well known as our society and do not have the same dominance. Why is this? It is because we are a society. They treat the society with deference because they all know that we are the way of the future."[9]

Pardess quickly ascended to a position of centrality in the Palestine citrus trade. In the 1912/13 season, for example, it shipped 174,000 cases to market—11 percent of Palestine citrus exports. The ascendancy of Pardess in Palestine citriculture at large and Jewish citrus in particular was due neither solely to the number of cases that it exported nor only to the size of its membership. Its strength, as noted above, traced to the way it was organized. Banks, foremost APB, gave it credit more readily than they would a private grower. Therefore, members of Pardess benefited not only from credit before the harvest and not only from collaborative selling but also from the advantages of a cooperative in the provision of "all materials needed for shipping of fruit, [treating] indications of tree disease, and chemical fertilizers." Pardess even developed new markets. Much of its fruit—large fruit from young orchards—was offered in the Istanbul market at this time, and at Pardess' initiative Palestine citrus was marketed to Russia (mainly via Odessa), Central Europe, and even Australia.[10]

It is also noteworthy that by the beginning of World War I, the success of Pardess had led to the establishment of two additional cooperatives, Union and Merkaz. Union was established in 1909, chiefly to counter the strength and dominance of Pardess. However, it succumbed to the longer-tenured cooperative and was forced to disband after only one season. Merkaz was founded several years later, also due to dissatisfaction with Pardess and, especially, with what the founders claimed to be the veteran cooperative's discrimination against those who newly joined. Indeed, most members of Pardess at the time came from Petah Tikva and were not eager to admit new members, especially from other moshavot. APB, as explained above, attempted in the year that Merkaz was established to unite all Jewish growers in one central cooperative. The failure of the unification, as both Pardess and Merkaz sources indicate, traced to the veteran cooperative's demand for preferential treatment in coordinating shipments and negotiating with the dealers. Its aim, in fact, was to make Merkaz its subordinate. World War I thwarted the continuation of unification attempts and thrust not only Pardess but all of Palestine citriculture into a grave crisis.[11]

The operational goals, the marketing method, and the strategy of Pardess remained essentially unchanged after Palestine came under British occupation and recovered from the ravages of the war. Pardess continued to seek new markets and expand old ones. It increased its shipments to areas outside the Liverpool exchange and developed additional markets in Glasgow, Hull, Belfast, and London. In 1925, Pardess became the first citrus exporter in Palestine to begin large-scale marketing to Germany. In 1930, Pardess' marketing destinations ranged from Liverpool to Canada and from Norway to Ceylon, Greece, and India. Pardess also continued to be the exclusive Palestine representative of shipping companies that moved the fruit to Europe. In 1930, it represented two British firms and one from Romania. Furthermore, Pardess continued to provide members with centralized service in purchasing packing material before the season and obtaining advances from banks on account of future shipments.[12]

Obviously, too, Pardess continued to perform the main activity for which it had been founded—marketing its members' fruit. Unlike the impression that its name (Pardess Cooperative Society until 1925; Pardess Cooperative Society of Orange Growers from 1925 on) and the description of its activities may give, until the middle of the 1930s marketing at Pardess was not cooperative but rather individual for each member. Unlike the California method, in which all members' fruit was taken to a central packinghouse and packed on a pooling basis according to its quality and size, at Pardess each grower did his or her own packing in his or her orchard. After packing, too, the cooperative handled each member's cases individually from the time they were taken to the Pardess warehouses at Jaffa port, or to the railroad station, up to their sale to wholesalers in the various markets. In the specific domain of marketing fruit, Pardess actually functioned as a sales agent for its members in return for a commission that would cover expenses that it incurred on their account. However, the members of Pardess marketed their fruit under stipulated trade names and in accordance with instructions from management, and shipping and picking dates were determined by the Pardess executive committee. Pardess marketed its members' fruit under a very small number of trade names and treated the safeguarding of their reputation as the foundation of its business and an essential basis for the high prices that it usually obtained. Contemporaneous testimonies indicate that until the early 1930s, when prices remained high, the markets were glut-free, and the number of Pardess members did not greatly exceed one hundred, all members wished to improve the sorting and packing of their fruit in order to maximize their return in the market. Hence, Pardess saw no need to adopt any marketing method other than the one it had practiced to date. Moreover, in the first half of the 1920s the higher returns obtained by sale "on the tree" as against sale by means of Pardess caused the Pardess membership to decline by about one-fourth (from 119 members to ninety).[13]

In the 1929/30 season, Jewish citriculture crossed a watershed in its move toward unification. The turnaround occurred at three levels: organizational pattern, marketing methods, and the acknowledgment that some form of unification was an outright necessity. As we will recall, 1929/30 was a poor season for all citriculture in Palestine. The export price declined from seven shillings "on the tree" to five shillings and the price of berara fell even more—to half of the 1928/29 price. The main reasons for the abrupt price decreases were a large crop from Spain, a 70 percent increase in exports of oranges from Palestine, and a high proportion of blemished or rotten fruit. Several factors explain the high incidence of rot and blemishes. They were not unique to that season only, but due to the large increase in harvests they led to dispiriting results that had not been known previously. The factors included the poor loading conditions at Jaffa port, vessels that were slow and poorly suited to shipping of fruit, and overloading of vessels. Most important among them, however, were the inferior quality of sorting and packing and surrender to the temptation to mingle berara and high-quality fruit in the same cases. The big losers that season were not the growers, whose large harvest compensated them for the decrease in prices, but the dealers. Ahead of the 1930/31 season, many dealers were no longer willing (or able) to buy fruit "on the tree."

Unsurprisingly, then, the members of Pardess emerged from the poor 1929/30 season largely unscathed. They received not five shillings per case "on the tree," like most growers, but six shillings. The Pardess report for that season reemphasized the importance of the cooperative method and stressed once again, in the society's opinion, the short-sightedness and low likelihood of success in "speculative" marketing.

The Pardess membership stopped declining in the middle of the 1920s, began to grow perceptibly in the 1926/27 season, and accelerated with even greater vigor after the poor 1929/30 season. Thus, in late 1930 Pardess had 133 members and thirty-two growers who marketed their fruit as membership candidates. A year later, in the summer of 1931, the numbers of members and candidates expanded to 158 and fifty, respectively. A year later, Pardess had 206 members and thirty-four candidates. During the last-mentioned season (1931/32), Pardess marketed 752,000 cases of citrus, 20.6 percent of all Palestine citrus taken to market.[14]

As the method of selling in Jewish citriculture changed, so did the organizational methods. In the early 1930s, a clear trend toward the formation of growers' cooperatives was evident and the proportion of unorganized Jewish growers declined. By the end of the period, 85 percent of Jewish citrus was being marketed cooperatively (as against 45 percent ten years earlier) and thirty-six growers' cooperatives had signed up with the Registrar of Cooperative Societies. Twelve of these cooperatives marketed the fruit of their members.[15]

The tendency toward cooperation among Jewish growers from the early 1930s on was not fortuitous. It was abetted primarily by the decline in prices and profitability, coupled with the increases in the population of exporting growers and harvests. It is also noteworthy that the formation of citrus growers' cooperatives was associated with another trend, the Mandate authorities' encouragement of this form of organization. One of the first actions of the British civil administration in Palestine was the promulgation of the Cooperative Society Ordinance in 1920. The main purpose of the statute was to facilitate the organization of any entity that aimed to promote its members' economic interests in accordance with cooperative principles. The cooperative articles of incorporation were amended in several ways during the Mandate era, but their basic goal, as set forth in 1920, did not change. Furthermore, in the early 1930s an attempt was made to organize rural Arabs into cooperatives following the Raiffeisen model—an experiment that did not work out well. With very few exceptions, only the Jewish sector took the cooperative path. Given the state of development of the capital market in the Jewish sector of Palestine at the time, cooperative organization was also a very advantageous way of obtaining credit.

As the 1930s began, plantings during the first "planting frenzy" (1926–1931) advanced toward productivity. The new growers in the Sharon area and the owners of the young orchards in Judea, who to this point had been preoccupied with matters related to preparation and initial cultivation of the soil, now faced the question of how to market the fruit. Many of them (those who owned young orchards only) had an especially daunting problem, because their fruit was large and, in most markets, less in demand than the smaller mature fruit. Moreover, the preparation and cultivation costs were much higher for Jewish orchards (especially the young ones) than for Arab-owned orchards. Thus, dealers—Arab and Jews alike—either offered a price for Jewish fruit that usually failed to cover growers' costs or refused to commit to a total price for each season, unlike their previous practice. The dealers even balked at allowing some of the harvesting, packing, and sorting to be done by parties unrelated to them, that is, by Jewish labor. In their method, a large group of pickers, packers, and sorters was taken to the orchard, set up a tent, and did the work quickly and cheaply. Jewish workers could not be integrated into this modus operandi because they demanded higher pay for picking and packing and, especially, refused to subordinate themselves to the dealers. In February 1935, Isaac Rokach addressed himself to this issue when he explained the main reason for the development of the cooperative form of organization in Jewish citriculture:

> Almost every Jewish grower in Palestine believes it necessary to join a cooperative because of the difficulty in finding a buyer at a satisfactory price. For obvious reasons, both Jewish and Arab dealers prefer to buy fruit from Arab orchards. It is not love of cooperatives that has forced the growers to join cooperatives, but necessity.[16]

The last phase in the changeover from individual marketing by dealers to cooperative marketing took place with the gradual adoption of the "price equalization" or "pooling" method. In pooling, a general average price was set and every grower received this price commensurate with the number of cases shipped. The average price was determined in various ways, usually in accordance with the time of shipment and the genus of the fruit. The pooling method was adopted due to the growth of the industry in terms of supply of fruit and the number of members of the cooperative. It was no longer possible to keep track of individual growers' cases and pay them for the exact quantities delivered.

Furthermore, the increase in early sales and the need to divide the proceeds among a large number of members made it necessary to establish a standard case in terms of genus and quality. By the same token, distrust among growers, which had a favorable effect on packing and sorting at times of limited supply and close individual inspection, now tempted growers to market fruit that was unfit for shipping along with high-quality produce.

However, acceptance of the pooling method and tough inspection of members' fruit at point of shipping did not suffice to solve the basic problem. Rotten and improperly sorted fruit continued to reach the markets. Truly effective inspection required a fundamental change in the sorting and packing method: (1) Growers would no longer pack their own fruit; (2) the use of small packing facilities in each orchard would be discontinued and central regional packinghouses would be built. This issue—revising the basic method of Jewish citriculture—was the main theme in discussions of the unification issue during the second part of the period discussed here.

Establishment of "The Jaffa Citrus Exchange"

Until 1929, as stated, Palestine citriculture dealt with the unification issue at the theoretical level only. It was clear to everyone that the conditions for such unification had not matured and that the only party that could impose it, the government, was trying not to interfere. In 1929, a significant change took place. The Arab uprising that year, which jolted Jewish public opinion in Palestine and in the Yishuv leadership and energized a process of social and economic separation of the Arab and the Jewish populations, did not skip over the citrus industry. From then on, it seems, conscious disengagement from the Arab sector became increasingly prevalent among Jewish growers and the gap between the two sectors of the industry widened steadily. The Jewish labor issue provides another clear reflection of this.[17] Pronouncedly economic factors combined with this consciousness and made their own contribution to the separation process, as we have seen.

For the purposes of our study, it is important to note that the separation process was manifested in additional aspects of Jewish citriculture activity, foremost marketing and marine transport. At the beginning of the period at issue, most fruit was marketed by Arab dealers who were affiliated with shipping companies. Even insofar as Pardess marketed citrus fruit, it did so in coordination with the other, non-Jewish, shippers. During the 1930s, in contrast, a nearly total disengagement occurred. Growers in each national sector contacted the shipping companies separately, and in many cases companies that served one sector did not serve the other.

The disengagement was also evident at the ports of departure. Haifa port did not open officially until October 1933 but citrus exports through Haifa—the first sixty thousand cases—began in the 1928/29 season. The users of the new port were Jewish growers, foremost those organized under the Pardess umbrella. Jaffa had become notorious for its poor loading arrangements; Haifa promised much more suitable conditions. It was a deep-water seaport where several vessels could dock concurrently at piers, thus obviating the need to subject the fruit to further rough handling by having it loaded onto small boats. Moreover, Haifa had direct rail connections with the citriculture areas. Theoretically, then, the fruit could be sent quickly and conveniently straight to the port, where a special citrus area has been set aside. Accordingly, the Pardess people believed in the early 1930s that they could ship their produce systematically and improve their inspection of packing and sorting by using Haifa port.

For this reason, Pardess decided to relocate its marketing center from Jaffa to Haifa. Since most Jewish shippers followed Pardess' lead, more than 90 percent of export shipments from the day the port opened until the end of the period at issue was Jewish. However, the proximity of Jaffa to the citrus areas of most of the old moshavot, along with various early mishaps in the use of Haifa port, led to some hesitancy about the changeover to Haifa as the port of departure until the 1936 Arab uprising, which stanched the doubts. Until the 1935 season, the proportion of shipping through Haifa did not exceed 40 percent of total exports from Palestine, whereas in the spring of 1936 it came to 60 percent, that is, most Jewish shipping.

Another result of the Arab uprising was the opening of Tel Aviv port, which strengthened the process of disengagement from Arab citriculture.

FIGURE 9.1. Jaffa Port—whereof most of Palestine citrus fruit was exported.

The proportion of Palestine citrus that was exported through Tel Aviv climbed from 3 percent in the 1936/37 season to nearly 10 percent a year later.[18]

The most salient expression of the disengagement of the national sectors in citriculture was the establishment of the Jaffa Citrus Exchange in 1929. The name of this enterprise, not coincidentally, was reminiscent of the California cartel. It was founded by a group of Jewish exporters due to difficulties created by Arab stevedores and exporters in Jaffa in exporting Jewish-grown fruit after the 1929 violence. The exchange was established by the three best-organized cooperatives and the leaders in citrus exports in the Jewish sector: Pardess, headed by Isaac Rokach; Jaffa Fruit Company, headed by Shmuel Tolkowsky; and Associated Orange Growers, headed by Zalman Jacobson and Lazar (Elazar) Rabinovitz.

Ostensibly and theoretically, the Jaffa Citrus Exchange was not closed to Arab shippers but de facto—obviously due to the circumstances underlying its establishment—it united Jewish growers only. During the ten years from its formation until the beginning of World War II, the exchange gradually grew and expanded. By September 1939, it included all Jewish fruit exporters, irrespective of groupings and ideological sectors.

Hopes were pinned on the exchange from its first day. Its founding members hoped that it would seize the opportunity and transform this voluntary organization not only into a place where growers of one national affiliation could identify and operate but also into a setting of cooperative economic activity. At first, when its membership was quite small, the exchange did organize marine transport contracts for its members and, pursuant to this, obtained several sizable discounts and improved the shipping terms. In 1935, however, as the membership grew and, mainly, "due to difficulty in reconciling the interests of several major shippers," as Tolkowsky put it, this collaborative activity ceased. From then on, the exchange was chiefly the Jewish growers' representative vis-à-vis the governments of Palestine and Britain; the various investigative commissions, to which it presented memoranda; and foreign governments. In the last-mentioned context, Shmuel Tolkowsky, chairman of the exchange, accomplished much near the end of the period—largely with semiofficial authorization from the Government of Palestine—to lower obstacles and barriers that had been raised against Palestine citriculture in countries that did not allow unrestrictive trade in citrus fruit. However, the exchange had no coercive power and its internal organization was voluntary. Personal rivalries among industry leaders and conflicts of interest among groups—for example, growers of young fruit versus those who produced more mature fruit, or private dealers incorporated within the exchange versus the cooperatives—prevented the Jaffa Citrus Exchange from unifying Jewish citriculture.[19]

Arab Exporters

Apart from disagreements and divergent interests among the Jewish growers, Arab citriculture was a perennial competitor with which cooperation was difficult to obtain. This was so despite efforts that were made at the beginning of the Mandate era and, subsequently, in meetings of the Citrus Committee and informal encounters. Economic interests that signaled to Arab exporters—as to many Jewish exporters—that it would not be worth their while to enter into the restraining framework of unification were among the decisive factors that thwarted closer cooperation.

This study has noted on several occasions the crucial role of Arab exporters, mainly large dealers and growers from the Jaffa and Ramle subdistricts, in the shipping of fruit. For most of the period of concern here, they controlled citrus shipping from Jaffa, the main export port for citrus fruit. Importantly, too, the Arab exporters also developed new markets, especially that of London, in which they were dominant—especially after the Jewish exporters began to market much of their fruit to continental Europe and after the Spanish exporters severely reduced their share in this market. The first Jewish entity to undermine the centrality of the Arab exporters was Pardess, back at the beginning of the century. However, Pardess never shipped more than about one-fourth of the citrus of Palestine and its share was usually smaller. Thus, until the early 1930s, with the establishment of the Jaffa Citrus Exchange, it was the Arab dealers who conducted most of the citrus trade. They financed the purchase of the fruit "on the tree" by means of advances that they received from brokers in England, and they regulated the shipping of fruit in accordance with demand.

The method used to regulate shipments was imperfect but not random. The exporters had little incentive to change it because it gave them high profits, especially when most of the fruit they marketed was cheaper to produce than the average Jewish fruit. The Arab exporters coordinated shipments at Jaffa port closely and rather effectively and were also in charge of apportioning storage space at the port in accordance with the exporters' national affiliations. In this range of activities—marketing the fruit, regulating shipments at the port, apportioning storage space, and attending joint meetings of the Citrus Committee—Jewish and Arab growers cooperated regularly. Furthermore, the Arab fruit dealers were organized under a national entity that paralleled the Jaffa Citrus Exchange, the National Chamber of Commerce.

By the end of the period, several groups of Arab growers and exporters controlled most of the marketed harvest in their sector, as did their Jewish counterparts in their respective sector. Among the Jews, in 1936 two main groups (Pardess and the Syndicate; see below) controlled 85 percent of the Jewish harvest, whereas among Arab exporters four groups controlled about 50 percent of the Arab harvest. The composition of Arab exporters and leaders of the Arab industry changed during the period. Here we pause briefly to describe several important personalities in the Arab industry in the late 1930s:

'Abd al-Rahman al-Taji of Wadi Hanin (Nes Tsiyyona) was a wealthy landowner from Ramle Subdistrict, a friend of Moshe Smilansky's, a well-to-do grower, and a senior member of the Citrus Committee and its subcommit-

tees. A relative of his, Shukri al-Taji of Jaffa, was both a grower and one of the four largest dealers in Arab citriculture at the end of the period.

Muhammad 'Abd al-Rahim, a wealthy notable in Jaffa and an important member of the Husayni party in Jaffa, controlled most Arab shipping of citrus during the Mandate era.

Alfred Rok belonged to a distinguished Christian family in Jaffa that claimed Crusader lineage and had orchards in the vicinity of Jaffa. During the Ottoman era, Rok had been accused of murdering a Jew from Rishon Lezion and was forced to flee the country. After he returned, he became one of the largest Arab fruit dealers, a member of the National Chamber of Commerce, a close associate of Haj Amin al-Husayni, and an activist in the Arab national movement.

Sa'id Beidas of Al-Shaykh Muwanis, a member of most of the citrus committees, owned considerable orchard dunamage. In 1940, he was named to the Citrus Marketing Board, which had only four members: Isaac Rokach, himself, and two government representatives.

Abdel Rauf al-Bitar was mayor of Jaffa, president of its Chamber of Commerce, and a member of the Citrus Inspection Board, which was established in late 1940.

Also noteworthy are the brothers *François and Antoine Jalant*, Christians from Jaffa. François Jalant, an orchard owner, member of the citrus committees, and member of the Citrus Inspection Board, attempted unsuccessfully to establish a cooperative of Arab growers.[20]

The Citrus Plantation Department

Another premature attempt to unify the Jewish citriculture industry was made in May 1931, this time at the initiative of the Farmers' Federation. On May 6, a "general assembly of growers," as the press called it, convened in Tel Aviv and resolved to establish a special department in the Federation, the Citrus Plantation Department, as an administration for the proposed unification. Places in the department management were reserved for all factions in Jewish citriculture—the "Labor citrus settlements," the commercial firms, and the private farmers—but the assembly made sure that members of the Federation would have an absolute majority. The purpose of the Citrus Plantation Department was to bring all Jewish growers, irrespective of factional affiliation, under one roof.

The unification initiative was powered by Moshe Smilansky and it was evidently enlightened by unification trends that seemed to have been successful upon the establishment of the Jaffa Citrus Exchange a year previously and the fact that just then (the spring of 1931), private farmers in the Sharon and Judea areas had come together for the first time under one organizational roof—the Farmers' Federation, headed by Moshe Smilansky. The champions of the idea warned about the dangers that the growers would face in the absence of unification and noted the poor 1929/30 season and the risk of spread of the "new disease," which appeared in 1930 and attacked the fiber of the sweet lemon rootstock—hazards that were still fresh in the Jewish growers' minds.

It quickly transpired, however, that most Jewish growers (about 75 percent) refused to join the department. Furthermore, neither the Histadrut nor the private exporters participated in the founding assembly of the department. In view of the clashing interests of the various factions in Jewish citriculture, the still-favorable market conditions (which had improved immeasurably in the 1930/31 season due to the cold snap in Spain and the small supply in Palestine), and the reluctance to unify under the auspices of the Farmers' Federation, the attempt failed. The Citrus Plantation Department redefined its goal. Instead of proposing to unify all Jewish growers, it began to concern itself mainly with scientific research, provision of pesticides and fertilizers, and collaboration with the experimental stations of the Zionist Organization and Mikve Yisrael.[21]

ISAAC ROKACH AND CENTRALIZATION OF PACKING

The attempts to unify Jewish citriculture that we have mentioned thus far took place overtly and were reported in the contemporaneous press. In contrast, when the general manager of Pardess, Isaac Rokach, attempted to revolutionize the internal organization of the Pardess cooperative—and, in turn, the citriculture industry at large—the matter did not reverberate in the public domain. However, the results of his experiment had far-reaching effects not only on Pardess but also on the structure of Jewish citriculture and, especially, the dynamics of the industry's unification process.

As noted, Isaac Rokach was the son of Shimon Rokach, the first director of Pardess, and the brother of Joseph Rokach, a grower and a member of Pardess like his father and his brother, and of Yisrael Rokach, subsequently the mayor of Tel Aviv and a minister in the Government of Israel representing the General Zionist Party. After Shimon Rokach's death in 1922, Pardess was placed under the directorship of three men, one of whom was his son, Isaac. When one of the directors (Dr. Weinberg) died in 1927 and the other (Meir Appelbaum) retired, Isaac Rokach remained alone as general manager. By then, he was already recognized as an important personality in Jewish citriculture, if not in Palestine citriculture at large. Under his stewardship, Pardess grew and became the leading entity in Palestine citriculture. It was under Rokach that Pardess moved assertively to tap new markets, expand old ones, and pilot the establishment of large, efficient packinghouses.

Rokach—insofar as one may deduce from his writings, articles, and personal diaries, and from contemporaries' testimonies—embodied a combination

FIGURE 9.2. Isaac Rokach.

of traits that included a well-developed business sense that had an occasional tough side, a sense of humor that did not lack for irony, a somewhat philosophical outlook on life, broad schooling, and command of languages. His attitude toward the Zionist Movement and the rapid development of the Yishuv was based, on the one hand, on a deep-seated Zionist consciousness that was nourished largely by Jewish and family traditions and, on the other hand, on understanding of and sincere human respect for Arabs and the Arab dealers, with whom he had been in daily contact since childhood. His innovative principled outlook on the future of Jewish citriculture seems to have drawn sustenance not only from belief in the power of "progress" in Western civilization, but also on a sober analysis of the socioeconomic realities of Jewish citriculture.

In the inaugural edition of *Hadar* (1928)—a monthly journal of which Rokach was a co-founder, editor, and leading spirit—Rokach presented his view of the future of citriculture in Palestine and, in so doing, articulated the basic principles of his worldview. As we recall, *Hadar* was meant, in terms of its format and most of its goals, to be the Palestine counterpart of the most important American citrus journal, *The California Citrograph*. Indeed, its original English title was *The Palestine Citrograph*. Rokach's journal made its debut in February 1928 and appeared regularly during most of the Mandate era. Always published in Hebrew and English, it was accessible to both the Mandate authorities and the Arab citrus growers and exporters. It was even distributed abroad.

The name, format, and goals of *Hadar*, and its editors' proposed solutions to difficulties that might beset the citrus industry, attest—despite the cautious phrasing used—to a worldview that was decidedly Western and innovative. The editors wished to base the industry not on the traditional model of Palestine citrus—cheap unskilled labor, lack of organizational centralization, and limited mechanization—but on the California model. The Pardess reports for 1927/28 and subsequent seasons, corresponding to Rokach's tenure in the management of the cooperative, spoke of the rising need, due to the expected increases in harvests and numbers of growers, to establish additional cooperatives and streamline packing by establishing several "efficient" central packinghouses.

By 1928, then, Isaac Rokach had evidently reached the conclusion that, for Jewish citriculture and especially for Pardess, it was essential to go over to centralized packing in order to maintain profitability under conditions of increasing numbers of industry players and supply.[22]

Our sources do not note when Rokach crafted his detailed plan for the reorganization of Pardess. Apparently, however, he took no action to implement the plan until the previous, premature efforts to unify the industry, in both of which he had been a partner, had failed. What did Rokach's scheme include? The main principle was *total centralization* under the Pardess umbrella. Within five years, members would no longer pack fruit in their own orchards. Instead, they would use regional central facilities. Thirty central packinghouses would be established for long-standing members; new members would have to build a central packinghouse at their own expense before they would be admitted to the cooperative. Above all, the growers would no longer be in total control of harvesting and packing; these tasks would be administered by the management of the cooperative, which would have the last say in any dispute. Until that time, growers hired and inspected their own pickers and packers. Now the management would call the shots in hiring and firing packers and inspecting all stages of harvesting, sorting, and packing. Everything related to packing would be run by a special committee headed by a manager of Pardess. The goal was to install the "pooling" method.[23]

The final goals of the plan were to standardize sorting and packing and to establish a standard case, so that Pardess' cases would have a better reputation and command higher prices than those of their rivals at a time of rising supply and falling profit expectations. The upshot of the plan, however, was the radical restructuring of Pardess, in which members would transfer the prerogatives of decision making and direct influence to the management.

Rokach's plan touched off fierce resistance in Pardess. This is unsurprising, since Pardess had eschewed cooperative marketing and defended growers' independence staunchly until then. What is more, many growers unaffiliated with Pardess continued to market their fruit "on the tree" and their profits had not yet been perceptibly harmed. Some growers, especially in Petah Tikva, accused Rokach of pursuing "imperialistic goals." Others acknowledged the necessity of the plan but counseled against "hasty" action. Rokach himself wrote that "conservative old men" were generating an "ill wind" at Pardess.[24] The general assembly of members, the institution that gave the members of the Pardess management their authority and that approved important decisions and appointments, rejected his scheme. The members of Pardess, it seems, were loath to forfeit their independence and had not internalized the need to make the marketing methods of their organization more cooperative.

The sources do not explain why Rokach maneuvered himself into a position where the endorsement of centralized packing became a condition for his continued tenure as general manager of Pardess. However, they do suggest that several factors acted in concert. First, Rokach seems to have feared for the fate of his cooperative, which his father had helped to found and consolidate, in the absence of a centralized packing program. This explains, to some extent, his insistence on having the Pardess institutions endorse his plan even if rejection would lead to his resignation. Second, there is no doubt whatsoever that Rokach was sincere in his view of the development of the industry. Evidently, however, the passion of his vision drove him to fulfill the vision even though the conditions for this had not yet matured. Third, the sources show that important activists in Pardess had hinted to Rokach that they would pass the program despite the strong opposition among rank-and-file members. When the crucial moment arrived, however, Pardess rejected the plan and Rokach felt that he had to resign. Finally, personal disagreements among members of the Pardess management and the personalities of several key players in the affair—foremost Rokach himself—helped to bring the question of Rokach's continued service to a boil.[25]

The Establishment of the Jaffa Orange Syndicate

Rokach tendered his resignation from the Pardess management in early 1933, although he continued to run the cooperative until the end of the season. His diaries show that he found the parting difficult: "The tenth to twelfth of this month," he wrote on April 13, 1933, "were harder for me than I had thought.

Hadn't I been accustoming myself to the idea of leaving Pardess for months? If so, why did I feel so sad that I could not hide my state of mind from others? Furthermore, hadn't I left Pardess of my own free will, even though dozens of members [of the cooperative] and the board, with Mr. Smilansky at their head, requested if not pleaded [that I remain]?" Rokach's separation from Pardess was painful and in fact, until he returned to the Pardess management as the general manager of Pardess Syndicate, under which Jewish citriculture was unified, he reserved a warm corner of his heart for the cooperative that his father had founded. Initially he remained a member of Pardess and continued to use the cooperative to market his fruit. However, several run-ins with the new management, all of which were business-related, evoked feelings of enmity and harsh rivalry on both sides. The altercations had a crucial effect on the industry's ability to function harmoniously in the years following Rokach's resignation.[26]

As noted above, Isaac Rokach did not abandon his involvement in the marketing of citrus after he resigned from Pardess. "As one who grew up in the orange trade," Rokach wrote to an important customer, "it would be only natural for me to remain in the industry after leaving Pardess."[27] By early 1933, during the lame-duck period between his resignation and the end of the season, Rokach had already begun to develop his new business activities. At his initiative, two new cooperatives, Adir and PFA (Palestine Fruit Association), were established in the Sharon area and he was chosen to head them. Concurrently, the Rokach brothers—Yisrael, Joseph, and Isaac—and other entrepreneurs formed the Jaffa Orange Syndicate, Ltd., a marketing company that held the exclusive right to sell the new cooperatives' produce. The syndicate itself also engaged in commission sale of citrus of growers, including Arab growers, who had not joined these two cooperatives. It quickly became the second-largest exporter in the Jewish citrus industry, after Pardess. Although Pardess never relinquished its lead, the establishment of the syndicate thrust the industry into competition over the "acquisition" of new members and customers. One of Rokach's most important achievements was the acquisition of Tolkowsky's Jaffa Fruit Company. In this transaction, the syndicate took over not only the company and its hundreds of thousands of export crates but also its trademarks, its relations with one of the large citrus trading companies in Britain, and one of its talented executives, Avraham Polani. Additionally, Rokach and several private exporters established a marine shipping company, SML, which competed with Pardess and the Arab forwarders for the organization of shipping and contracting with shipping companies.

Rokach's resignation from Pardess accelerated the formation of cooperatives in Jewish citriculture. From then on, Pardess displayed greater willingness and alacrity in admitting new members. Furthermore, as the syndicate was being set up in 1933, several large growers in Petah Tikva established a cooperative of their own, Bustan. Bustan, like the cooperatives that Rokach ran, used the pooling method and applied Rokach's centralized-packing scheme, albeit on a small scale. About five years later, in 1938, Bustan joined the syndicate.

It deserves emphasis that the process of incorporation of Jewish citriculture, as explained above, was driven by obvious economic factors and did not base itself on the initiative of anyone in particular. However, the enlistment of a powerhouse such as Rokach in the new citrus cooperatives and his support of cooperatives that favored centralized packing, such as Bustan, led to two clashing results. It expedited the transition to fully cooperative marketing; in this sense, Rokach's resignation abetted the unification of Jewish citriculture. However, it also stirred up superfluous competition within Jewish citriculture and caused direct detriment to profitability. Moreover, Rokach and the new management of Pardess very quickly developed relations of enmity and distrust. These turgid relations did not help to hasten the unification of Jewish citriculture when the years of crisis arrived.

THE FORMATION OF PARDESS SYNDICATE

The industry leaders often cited the downslide in relations between the two large private-citriculture entities and the escalating rivalry between them in markets, marine shipping, and enlisting of new members as the greatest obstacle to the unification of the Jewish industry. From the standpoint of Smilansky, chairman of Pardess and president of the Farmers' Federation at the time, the worsening economic conditions and the upturn in struggle between Rokach and his erstwhile colleagues obviated the possibility of unifying the industry on the basis of trust and reciprocity, that is, an alliance among the cooperatives. Therefore, in the mid-1930s he proposed "not a unification of cooperatives but one unified cooperative." This cooperative, he said, should be based on the "centralization method," meaning that "selling, buying, and shipping should be centralized." Although his analysis of the state of Jewish citriculture was sober and thoroughgoing, Smilansky's practical proposals reverted to the same old format, which assured Pardess control of the unified entity without reorganization. In Smilansky's united-cooperative scheme, the existing cooperatives would continue to exist. Furthermore, they would continue to arrange credit for their members and, above all, would organize "matters related to picking and packing, inspection of fruit quality, [and] apportionment of shipments among members." The upshot of the plan was that Pardess would control the new entity, since it still controlled 50 percent of the Jewish citrus crop. By implication, too, even as the syndicate used the pooling method and aimed to standardize marketing, Pardess would continue to ship fruit that was packed in nonstandard ways and roll the losses occasioned by this onto the other members of the proposed entity.[28]

Between 1933 and 1935, then, the Pardess management seemed to change its attitude in that it acknowledged the need to dissolve the various cooperatives and establish one central entity. However, it could not accept the idea of self-liquidation; instead, the veteran cooperative proposed that the liquidation include all cooperatives save Pardess. Pardess would "swallow up all the other cooperatives," Rokach explained, and would head the proposed united entity without making significant changes in the sorting and packing method. For Rokach and the members of the other cooperatives, this meant returning to the Pardess way of packing and sorting and the loss of the other growers' independence, all for no real gain to them. Their rejection of the Pardess proposals was inevitable. The crisis in Jewish citriculture had apparently not risen to a level that would entail a fundamental change in thinking. Such a process is typical of such situations, as Mancur Olson pointed out. Be this as it may, the end result was the perpetuation of the status quo.

The 1936/37 season was *the* year of crisis in Palestine citriculture at large and in Jewish citriculture in particular. For the first time, the growers at large, not only those who owned young orchards, failed to record a profit, and all growers of grapefruit sustained losses. Furthermore, due to the expansion of planted area, the high standard of living that had been established during the prosperity era, and the decline in prices, Jewish citriculture, in the estimation of an expert such as Smilansky, ran up P£4 million in debts. Only one-fourth of the Jewish growers were debt-free; all others were indebted to some extent, in many cases beyond the value of their orchards. Another expert, Harry Viteles, estimated the debts of Jewish citriculture more cautiously, at a nevertheless hefty P£2.85 million. Notably, in Viteles's estimation, the Jewish growers incurred only 17 percent of the indebtedness of Palestine citriculture all told.[29] All this aside, the menace of the burgeoning harvest loomed on the horizon continually.

In this state of affairs, rank-and-file growers also began to speak out in favor of unification. At Pardess, despondency and gloom seem to have become the order of the day. The management began to face criticism over outsized personal expenses, unnecessary investments in too many properties that were too fancy and in miscellaneous materials and equipment, faulty marketing, and an absence of sincere willingness to unify the industry. It does seem true that Pardess was less willing than the other cooperatives, psychologically and organizationally, to make the transition to unity. In 1938, a large majority of its eight hundred members still packed fruit the old-fashioned way, using small packing facilities in their orchards. Above all, as the critics said and as we showed above, the leaders of Pardess still entertained hopes of maintaining their cooperative's domination of the Jewish citrus market.

The results of the 1936/37 season, however, forced even the Pardess people to take action and make what they considered painful concessions.[30] The process seems to have begun at a secret meeting between executives of the large banks, who were alarmed about the state of the industry, and Jewish citriculture leaders, at King David Hotel in Jerusalem on March 1, 1937, even before the season ended. The bankers established a special consortium, which they called the "orange consortium," to deal with the citrus crisis. The questions that hovered in the background, which they also asked overtly, were whether Jewish citriculture had the strength to survive the 1936/37 crisis and what the growers intended to do to fortify the industry.

Importantly, the Jewish cooperatives marketed 85 percent of Jewish growers' fruit at the time. Consequently, any question about the strength of Jewish citriculture was actually a question about the strength of the cooperatives.

At the meeting, the growers attempted to toss the ball into the court of the Palestine Government and to use the banks as their advocates. It was indeed important to change the government's attitude toward citriculture in respect to taxation, tariffs, and other matters. However, the bankers did not fall for this tactic. Instead, they subjected the unsuitable structure of the industry and its leaders' unwillingness to effect the unification to fierce if not withering criticism. Again they lectured the growers about the advantages of unification in cutting costs and improving return in the market. Summing up the meeting, the "orange consortium" voted confidence in the strength of the cooperatives but urged them to unite.

Thus, in the summer of 1937 the Pardess general assembly adopted an important resolution in regard to reorganization. The resolution gave the cooperative's branch offices greater autonomy and instructed them to operate in accordance with the articles of local cooperatives. The reconfigured Pardess would open its ranks to all comers and become "the central institution of all Jewish citriculture." Joining Pardess would no longer "resemble an act of 'entry' but participation on the basis of equal rights."

Predictably, Rokach and the Syndicate people did not take well to this approach. At a conference of pro-Syndicate growers in Tel Aviv, prominently covered in *Hadar*, the managements of the cooperatives came under ferocious criticism for "holding back the unification." Rokach, in a review of the discussions in advance of unification, depicted the Pardess proposal as another way in which the veteran cooperative would propose a "full merger" of the two cooperatives. The syndicate, Rokach said, did not believe it could accept the proposal but tabled a counterproposal of its own: the conclusion of notarized contracts between the cooperatives that would institute formal cooperation in limited fields. Rokach's pressure seems to have paid off; Pardess withdrew the merger proposal. However, no consensus was reached by the end of 1937 and the preexisting organizational structure was used to market the 1937/38 harvest. Toward 1938, however, the sides accepted a formula at the level of principle: Each cooperative would remain administratively autonomous but the unified entity would wield supreme authority in matters of packing and sale of fruit, purchase of materials, and marine shipping.

At this stage, several issues surfaced that threatened to lead the negotiations to an impasse. The first concerned the role of the two Labor Movement cooperatives, Yakhin and Tnuva Export, in the proposed unified organization. The second issue pertained to the extent of the parties' psychological willingness to be unified, and the third issue had to do with management and the managers. In March 1938, these questions boiled into a crisis.

The Histadrut-affiliated cooperatives demanded that the unified entity hire only, or at least mainly, Jewish labor. The members of "unaffiliated" cooperatives considered this an obstacle to the cost cutting that the unification was supposed to achieve. As for the psychological willingness and the ability to disengage from the erstwhile organizational patterns, it seemed that Pardess, in particular, found the matter difficult. Its March 1938 proposal concerning the structure of the unified body and the prerogatives of its directors and managers, as submitted to Rokach and mentioned at length in his diary, attests to this. The Pardess proposal had been made, he claimed, "in the thought that the whole unification would be in effect for three years only" and the "organizers" were already contemplating the possibility of "having to leave even before the three years are up." The main bone of contention between Rokach and Pardess apparently had to do with the management issue. Rokach had not accepted the demand for two equally empowered managers when it had first been brought up in 1935 and opposed it now as well.

In fact, however, Rokach was very much of two minds. On the one hand, it was clear to him that "without unification there may already be disaster by the coming season." On the other hand, he found Pardess' terms very hard to accept.

> My dilemma is this: if I don't help to pass the unification under these conditions, there will be no unification, at least this year, and the public at large will blame the Syndicate and me. We'll have poor results next season. Who's at fault? The syndicate and the man who heads it, Mr. Rokach. If I help and the unification goes through—under the current conditions—I risk my reputation, the Syndicate, and the entire unification.[31]

At this point, as the debates over unification slid into deadlock and the economic and political situation in Europe subjected Palestine citriculture to rising difficulties, the banks intervened. Several days after the negotiations ran aground, Viteles wrote to Zalman Jacobson, who at the time was one of the main personalities in Pardess and someone whom even Rokach accepted, and proposed that an outside player help solve the unification problem. In early May 1938, with the intervention of Julius Simon (a leading figure in the Brandeis group and president of the Palestine Economic Corporation, which the group had formed), a committee of three bankers was established to "make proposals in regard to establishing the unification." The members of the committee were Eliezer Hoofien (chair), general manager of the Anglo-Palestine Bank, Harry Viteles, general manager of Central Bank for Cooperative Institutions, and Mark Yaffe, a member of the administration of the influential and less-known investment corporation (which was based in London), The Palestine Corporation.

The "Committee of Three" acted swiftly; on July 15, 1938, it released a "platform for the unification of the citrus growers' organizations." On September 6, 1938, all Jewish cooperatives endorsed the platform and appointed a small committee to iron out several disputed points. On December 15, the Committee of Three republished the platform in an amended version. Its main provisions were the following: (1) The unification would centralize "no less than 80 percent of the entire Jewish citrus harvest." This meant that Yakhin and Tnuva Export would have to be included. (2) The unification would be time-unlimited and in any case would not be liquidated until at least five years after the date of its founding. (3) The large existing institutions of the syndicate and Pardess would be dismantled. (4) Pardess' branch offices would become, and would be registered as, "autonomous cooperative associations" within two years of the founding of the unified entity. (5) As for the question of the managers, the committee made no firm stipulations but proposed that Ellman of Pardess and Rokach of the syndicate serve as "general managers with equal status." (6) The founders of the unified institution would assign their trademarks to this entity without compensation and there would be no trademarks save those determined by the unified institution. (7) Growers would no longer be responsible for packing and sorting; the unified entity would be empowered to hire packers and sorters and be the sole payer of their wages. (8) The unified institution would be the sole decision maker in matters of shipping and haulage.[32]

After the amended platform was released, the various cooperatives approved it with celerity. Thus, the establishment of Pardess Syndicate was announced on February 12, 1939, although the industry finished the 1938/39 season on the basis of its former modus vivendi. The Registrar of Cooperative Societies recorded the new cooperative officially and finally on April 18, 1939. However, the formation of Pardess Syndicate did not mark the completion of unification as the Committee of Three had expected. Pardess Syndicate encompassed not 80 percent of the Jewish citrus harvest, as the committee had stipulated, but 60 percent. The main reason was the nonenlistment of Yakhin and Tnuva Export. These two cooperatives controlled 1.8 million export crates in the 1938/39 season, 18 percent of marketable Jewish citrus, a quantity similar to that exported by the syndicate that year. Had these two organizations joined, Pardess Syndicate would have met its 80 percent target.

Yaknin and Tnuva Export stood aside due to two main issues: Jewish labor and the composition of the management. The former problem was solved in the course of discussions after the first version of the unification platform was accepted, when Smilansky stated that Pardess Syndicate would accept 80 percent Jewish labor in picking and packing. The remaining issue concerned representation in the management. The Labor cooperatives demanded representation commensurate with their rising strength and, more important, sought to keep the management from amassing unchallengeable coercive power. The non-Labor cooperatives, in contrast, wished to strengthen the management and empower it generously. Furthermore, the view was expressed that "the man in the citriculture street" did not trust the Histadrut and its representatives and feared that the Laborites were scheming to take over the new organization. Thus, the non-Labor cooperatives neither wished to bring Yakhin and Tnuva Export into the unified entity nor saw no possibility of so doing.

Viteles and Hoofien accused Smilansky, Rokach, and the others of rushing to announce the unification and failing to give the Histadrut people a practical chance to work a compromise. The non-Labor cooperatives rejected a proposal from the Labor cooperatives that the Committee of Three arbitrate between them and the other cooperatives on the issues that had arisen between the sides. Hoofien, in a letter to Smilansky several days after the unification was officially announced, expressed his impression that the non-Labor cooperatives were no longer interested in the intervention of the Committee of Three. Hoofien did not respond to Smilansky's invitation of the members of the committee to the festive inaugural meeting of the Pardess Syndicate board. The main rationale for the rejection was the committee members' disappointment over the share of exported fruit that the Pardess Syndicate had gathered, less than the 80 percent proportion that they had stipulated in their initial proposal.

There were additional reasons for the partial success of the unification. Many members of Pardess seceded from their cooperative and did not join Pardess Syndicate. A year before the unification resolution, in early 1938, a large and long-standing group of Nes Tsiyyona growers broke away from Pardess and established Pri-Taz. They championed the old methods of Pardess in regard to packing, sorting, and Jewish labor. Accordingly, they found it difficult to accept Smilansky's statement in regard to 80 percent Jewish labor and the centralized packing issue. In addition to Pri-Taz, which marketed about 400,000 crates in the 1938/39 season, members whose total exports added up to a similar quantity seceded from the cooperatives that had formed Pardess Syndicate. Together, the breakaways accounted for 9 percent of Jewish citrus exports that season.

The party that seems to have paid the highest price for unification was Pardess. Even though it remained the largest player in Jewish citriculture, its relative strength had declined in one stroke. It had to disintegrate into autonomous regional cooperatives and its members accepted the principle of centralized packing, which they had opposed so vehemently in past years and that had prompted Rokach to resign from the cooperative only six years previously.

Indeed, the establishment of Pardess Syndicate was Isaac Rokach's victory. In the final form of the unification, Rokach and his associates held a majority on the board of directors, even though their organization was smaller than that of their rivals. Rokach also got his way in regard to the composition of the management. Instead of two equally empowered managers, one general manager for Pardess Syndicate was appointed, and the post went to none other than Isaac Rokach. Rokach triumphed in the matter of packing; his demand that packing be placed under the central control of the Pardess Syndicate management was accepted. Furthermore, the central cooperative was the only entity allowed to sign contracts for the haulage of fruit. Finally, sole authority for the packing itself was reassigned from the growers to the cooperatives.

The 1939/40 season was the most bounteous one that Palestine had ever known, with more than fifteen million crates exported. However, it was both the first and the last season in which Pardess Syndicate operated as a marketer in a free market. World War II, which began in September 1939, threatened to damage the entire industry. As it soon turned out, however, Mediterranean transport would not come under restrictions in the winter and spring of 1940. In advance of the season, Pardess Syndicate concluded agreements with most Jewish marketers to limit sales of grapefruit and to coordinate sales of oranges in countries that did not allow free trade in citrus fruit. Furthermore, when the war began, the new cooperative proposed severe limits on the quantity marketed from Palestine in order to improve profitability.

However, as Rokach explained at the end of the season, "shippers in both groups, the Jewish and the Arab, could not work out an agreement over the terms, and government did not wish to force the shippers to accept any imposed agreement. . . . After weeks of negotiations, the emergency marketing plan was shelved and the ordinary disorder, with all its wretched results, set in. The results were even more wretched this season due to higher expenses, especially for marine shipping." Even the agreements among Jewish growers were not honored, and the familiar phenomenon of destructive competition that drove prices below the previous year's levels recurred, Rokach said. The situation worsened further due to difficulties that the growers had encountered in the spring of 1939 in obtaining advances to cultivate their orchards and prepare for the marketing season. Consequently, many orchards were not cultivated on time.

As stated, 1939/40 was the last season in which Palestine citriculture functioned as an arena of free enterprise. Shortly after the end of the season, the Government of Palestine changed its policy and placed the Palestine economy on a war footing. By the summer of 1940, it was clear that the government had decided to put the citriculture house in order. Two citrus boards, one for inspection ("The Citrus Control Board") and the other for marketing ("The Citrus Marketing board"), came into being in December of that year and have regulated marketing and other industry affairs not only until the end of the Mandate in 1948, but long after the establishment of the State of Israel.[33]

TEN

Conclusion

Jewish Citriculture as a Private-Enterprise Industry

THIS STUDY HAS PRODUCED a portrayal of citriculture, an important private-enterprise industry in the Yishuv (the nascent Jewish community of pre-independence Israel), from several angles: the ideological views and social profile of its players, its spatial spread, and the way it conducted its affairs.

From the ideological standpoint, we found that the private growers had a cohesive worldview in which the individual and the principle of "natural rights"—personal freedom and the right to property above all—underlay a comprehensive socioeconomic method that, they believed, would lead not only to personal welfare and, in turn, the welfare of society at large, but also to the fastest and most efficient attainment of national goals. In other words, they saw no contradiction between the liberal capitalistic worldview, to which they subscribed, and a national worldview, and even claimed that their worldview itself would lead with the greatest alacrity and efficiency to the attainment of the national aims—the creation of a Jewish majority in Palestine, the acquisition of most of the country's land, and the development of economic and social strength for its Jewish inhabitants.

As for spatial distribution and social profile, we found that Jewish citriculture in Palestine traced its genesis to capitalists of considerable means and was concentrated mainly in one part of the country, that known as Judea in the jargon of the time. During the period at issue, however, the industry spread to most areas of Jewish settlement and most Jewish farming communities, private and nonprivate alike, made it part of their agriculture. Still, the greater part of the industry remained in the hands of private capitalists, some rural and others city-dwelling. Thus, they were private entrepreneurs and most economic activity in citriculture was private enterprise.

At the beginning of our discussion, we defined several characteristics of the entrepreneur, distinguished between private and public entrepreneurship, and listed several noneconomic factors that affected the extent of entrepreneurial activity in the Yishuv society. We found various types of entrepreneurs among the growers. Some belonged to the Schumpeterian mold: They blazed new trails, changed the equilibrium, and introduced material changes in organization and economic activity. The most prominent of them were Shimon Rokeach and his son Isaac, who may be described as the premier players in the industry during its two eras of development, the Ottoman and the Mandatory; Moshe Smilansky, the leader and innovator, especially in respect to the organizing of growers and farmers; Oved Ben-Ami and his associates at Hanotea who founded and developed Netanya; and Yehuda Horin, the public entrepreneur who ran the Histadrut contracting firm Yakhin. All not only created new paths but also identified opportunities where others saw only uncertainty. Most rank-and-file growers, however—the thousands of individuals who invested in the plantation and cultivation of citrus orchards and the marketing of their yield—were private entrepreneurs who did not blaze new trails but acted and took risks under conditions of uncertainty.

What noneconomic factors affected the extent of private entrepreneurial activity in citriculture? The first was the legitimacy of private enterprise in this industry, which the political and ideological "center" of the Zionist Movement and the Yishuv, as Horowitz and Lissak defined it, circumscribed rather severely. After all, the growers offered a strong alternative to the center's actions in some of its most important symbolic fields: agricultural settlement, return to the soil and manual labor, ownership of land, social justice, and Jewish labor. In the main, those in the center were concerned that the positioning of individual utility—which derives its nourishment from the essence of private enterprise—over collective utility would cause the national and social goals of the agricultural settlement enterprise to be neglected and would lead to "exploitation" of the Jewish worker, the sale of land to Arabs, and the employment of Arabs in the Jewish economy. The growers, in turn, alleged that political intervention in the system of economic decision making would impair the autonomy and economic efficiency of the system and disserve the very national and social goals for which the political mechanism intervened in the growers' free economic activity in the first place. The struggle over the Arab labor that was used in citriculture, the demand for non-"exploitation" of the Jewish worker, and the derogatory portrayal of private ownership of land were regular features of the contemporaneous culture that kept Jewish private enterprise in a permanent state of weak legitimation and delegitimation. This atmosphere, however, did not stop the private enterprise from expanding and developing.

Two main factors allowed private enterprise to continue functioning at this time. First, as Alexander Gershenkron noted, the negation of entrepreneurship by society and the carriers of its values does not suffice to bring entrepreneurial activity to halt; such an outcome requires government support. During the period at issue, however, and especially under the British Mandate, the government of Palestine acted to encourage private enterprise because it shared the growers' worldview. Indeed, even the growers affirmed government action in the domains that are "classically" reserved for the public sector—legislation and its enforcement, infrastructure investment, delivery of services (such as security and defense), and research and development. In these regards, not only the Mandatory Government played the role of a public sector from the growers' perspective; so did the official Zionist institutions. However, as stated, the growers rejected attempts by the Zionist Movement and its labor movement to interfere with unrestricted economic activity and to steer this activity in the direction of their goals.

The second factor that abetted private enterprise was the continuity of the growers' operations relative to their economic activity before they reached Palestine. Nearly all citrus growers had been small- or medium-scale capitalists in their countries of origin. Although few had personally been farmers before they came to Palestine, their transition to citriculture did not require them to deal personally with physical cultivation. Many growers did not even settle on their land, let alone cultivate it. In other words, the growers' activity, both in the Diaspora and in Palestine, was primarily capitalistic. Most growers, however, chose to live in areas of agricultural settlement, giving further evidence that many of them had assimilated the Zionist value system, which placed agricultural settlement on one of the highest rungs of its ladder of ideological preferences. Indeed, citriculture, as contemporaneous sources frequently note, was one of the types of agriculture that allowed urban Diaspora Jews to integrate very easily into the Zionist symbolic and social matrix in Palestine.

As for their status in political and symbolic-values terms, the private growers were on the fringes of the center. From the economic and social standpoints, however, private citriculture was definitely an "alternative center." Many growers even took part in operations of the Hagana (the military organization of the center). Arguably, then, the various elements in center-periphery relations canceled each other out, leaving only the economic factor to determine the extent of the growers' entrepreneurial activity. Indeed, it was the economic factor, more than any other, that helped to change the attitude of the Zionist labor movement itself toward the free market and the legitimacy of activity in it. The movement's recourse to the market and its mechanisms, however, did not engender a favorable principled attitude toward private enterprise, since, after all, it left the individual and not the collective and its "will," at the center of its universe.

In respect to economic activity itself, we found that once growers decided to invest in a citrus orchard in Palestine, the chief motive in their actions was the maximization of profit. This goal was heavily influenced by the general nature of the new Jewish settlement enterprise in Palestine as a venture engineered (in the main) by Europeans who embraced modern culture, acted in modern economic ways, and even brought considerable capital. The orchard investment generated a handsome return that the Jewish growers attempted to increase even more by making various technological improvements. They did this to cut costs, on the one hand, and to increase yield per land unit and enable the investment to mature more rapidly, on the other. Indeed, the Jewish growers boosted the yields of their orchards to levels that exceeded those of traditional orchards many times over, and they reduced the time to first yield by about three years relative to the norm in the indigenous citriculture. However, the availability of factor inputs—land, labor, and capital—had a crucial influence on the extent to which Palestine citriculture adopted technological improvements relative to the universally accepted ideal, the California model. Jewish citriculture, then, fit into the space between the California model and the traditional Arab one. Additional factors were the ideological and conceptual elements of disengaging from the Arab economy, on the one hand, and modernization, which played a considerable role in the growers' economic activity and resource allocation for it.

Arab citriculture coexisted with Jewish citriculture and developed at a similar pace and to a similar extent. Our study found that the Jewish settlement enterprise—especially in its land purchases and the example that it set—inspired more than a few rural Arabs to take up citriculture and led to perceptible modernization processes in the Arab village. We also found that even though the two national sectors of the industry attempted to cooperate, a socio-national schism came into being upon the establishment of the Jewish sector and widened steadily as time passed.

Thus, our discussion of citriculture has opened a "window" through which we may observe economic realities, socio-organizational patterns, and ideological outlooks that were current in much of the Jewish society of pre-independence Israel, that offered a genuine alternative to the reality and ideology that the center attempted to dictate, and that, in fact, outlived the British Mandate.

Notes

CHAPTER ONE. THE TANTALYZING AROMA OF CITRUS BLOSSOMS

1. Anita Shapira, *Futile Struggle: The Jewish Labor Controversy, 1929–1939* (Tel Aviv: Hakibbutz Hameuchad, 1977), 53 (Hebrew); Idem., *Berl Katznelson: A Biography* (Tel Aviv: Am Oved, 1983), 223 (Hebrew).

2. Shapira, *Futile Struggle*, 214–352. On the cultural *Streit* between these camps, as well as on the construction of the opposing images of each camp, see also: Hannan Hever, *Poets and Bullies: The Evolution of the Hebrew Political Poetry in Palestine* (Jerusalem: The Bialik Institute, 1994) (Hebrew); Oz Almog, *The Sabra: The Creation of the New Jew* (Berkeley: University of California Press, 2000).

3. For a few representative studies of the "establishment" school see: Alex Bein, *History of Zionist Settlement: From Herzl to the Present* (Ramat-Gan: Massada, 1970) (Hebrew); Chaim Gvati, *A Century of Settlement: The History of Jewish Agricultural Settlement in Eretz-Israel* (Tel Aviv: Hakibbutz Hameuchad, 1981) (Hebrew). On the place of the private sector in general and private enterprise in particular in the economy of the Yishuv see, for example: Nachum T. Gross, "Private and Public Enterprise in Building Jewish Palestine According to Z. D. Levonteen," *Economic Quarterly* 30 (1983): 488–94 (Hebrew): Idem., "Free Enterprise Versus Planned Markets," in Nachum T. Gross, *Not by Spirit Alone: Studies in the Economic History of Modern Palestine and Israel* (Jerusalem: Magnes Press and Yad Izhak Ben-Zvi, 1999), 355–65 (Hebrew); Dan Giladi, "The Private Farmers' Stand towards the Issue of 'Hebrew Work' in the Moshavot, 1904–1914," *Baderech* 6 (1971): 69–73 (Hebrew); Idem., *Jewish Palestine during the Fourth Aliya Period (1924–29): Economic and Social Aspects* (Tel Aviv: Am Oved, 1973) (Hebrew); Hagit Lavsky, "Businessmen, Zionism, and the Building of the National Home: The Case of Salman Schocken," in *Jews in Economic Life*, ed. Nachum T. Gross (Jerusalem: Zalman Shazar Center, 1985), 357–72 (Hebrew); Idem., "Has There Been an 'Essential' Connection between Jews and Capitalism?: The Case of Mandatory Palestine," in *Religion and Economy: Connections and Interactions*, ed. Menahem Ben-Sasson (Jerusalem: Zalman Shazar Center for Jewish History, 1995), 387–400 (Hebrew); Idem., *Before Catastrophe: The Distinctive Path of*

German Zionism (Detroit and Jerusalem: Wayne State University Press, The Magnes Press, Leo Baeck Institute, Jeruslaem, 1998); Jacob Metzer, *National Capital for a National Home 1919–1921* (Jerusalem: Yad Izhak Ben-Zvi, 1979) (Hebrew); Idem., *The Divided Economy of Mandatory Palestine* (Cambridge: Cambridge University Press, 1998); Yaacov Shavit, *The Textiles in Eretz Israel 1854–1956: From a Pioneering to a Leading Industry* (Tel Aviv: Israel Textile Association, 1992) (Hebrew); Yossi Beilin, *Roots of Israeli Industry* (Jerusalem: Keter, 1987) (Hebrew); Yossi Katz, *Ye Shall Grant a Redemption for the Land: The "Geula" Society for Land Acquisition 1902–1914* (Jeruslaem: Yad Izhak Ben-Zvi, 1987) (Hebrew); Idem., "Private Zionist Initiative and the Settlement Enterprise in Eretz-Israel in the Early 1900's: 'Nationalistic Capitalism' of Private Capital," in *The Land that Became Israel; Studies in Historical Geography*, ed. Ruth Kark (New Haven, London and Jerusalem: Yale University Press and The Magnes Press, 1989), 275–86; Idem., *Zionist Private Enterprise in the Building of Eretz-Israel during the Second Aliyah 1904–1914* (Ramat Gan: Bar-Ilan University Press, 1989) (Hebrew); Irit Amit, "The Place of Great Capital in Shaping the Sharon in the Second Half of the Second Decade of the Twentieth Century," *Studies in the Geography of Israel* 14 (1993): 122–41 (Hebrew); Idem., *Jewish Settlement Activity in the Southern Sharon 1918–1929: National Capital and Private Enterprise in the Shaping of the Southern Sharon* (Unpublished PhD dissertation, the Hebrew University of Jerusalem, Jerusalem, 1993) (Hebrew); Idem. "Private Jewish Investors in Palestine in the 1920s," in *The Mosaic of Israeli Geography*, ed. Yehuda Grados and Gabriel Lipshitz (Beersheva: Ben-Gurion University of the Negev Press, 1996), 451–59; Yakir Plessner, *The Political Economy of Israel: From Ideology to Stagnation* (Albany: State University of New York Press, 1994).

4. Another issue, which stands beyond the scope of this study, is the gap that existed between the private sector's primary role in the economic and social life of the Yishuv and its weak stand in the Yishuv's political arena. See: Dan Horowitz and Moshe Lissak, *Origins of the Israeli Polity: Palestine under the Mandate* (Chicago: University of Chicago Press, 1978); Giladi, *Jewish Palestine during the Fourth Aliya*, 256–64; Yaacov Shavit, "The Jewish Community in Eretz Israel in the Mandatory Period: Between a Society and Polity," *Cathedra* 14 (1980) 7–16 (Hebrew); Zeev Sternhell, *Nation-Building or a New Society?: The Zionist Labor Movement (1904–1940) and the Origins of Israel* (Tel Aviv: Am Oved, 1995) (Hebrew); Anita Shapira, "The Politics of the Jewish Community in Palestine, 1918–1939," in *The History of the Jewish Community in Eretz-Israel since 1881: The Period of the British Mandate, Part Two*, ed. Moshe Lissak, Anita Shapira, and Gavriel Cohen (Jerusalem: The Israel Academy for Sciences and Humanities, The Bialik Institute, 1994), 1–70 (Hebrew); Idem., "Sternhell's Complaint: 'Nation Building or a New Society' by Zeev Strenhell," *Yiunim Bitkumat Israel* 6 (1996): 553–67 (Hebrew); Yonathan Shapiro, *The Historical Ahdut ha-'Avoda: The Power of Political Organization* (Tel Aviv: Am Oved, 1975) (Hebrew); Amir Ben-Porat, *The Bourgoise: The History of the Israeli Bourgoises* (Jerusalem: Magnes Press, 1999) (Hebrew).

5. Jacob Metzer and Oded Kaplan, *The Jewish and Arab Economies in Mandatory Palestine: Product, Employment, and Growth* (Jerusalem: Maurice Falk Institute for Economic Research in Israel, 1990), 102, 109–19, 159–60 (Hebrew); Michael Beenstock, Jacob Metzer, and Sanny Ziv, "Immigration and Jewish Economy in Mandatory

Palestine," *Research in Economic History* 15 (1995): 149–213; Nachum T. Gross and Jacob Metzer, "Public Finance in the Jewish Economy in Interwar Palestine," *Research in Economic History* 3 (1978): 87–159; Nadav Halevi, "The Economic Development of the Jewish Community in Palestine 1917–1947" (Jerusalem: Maurice Falk Institute for Economic Research in Israel, Discussion Papers No. 7914, 1979) (Hebrew); Abraham Ulitzur, *National Capital and the Building of the Country, 1918–1937* (Jerusalem: Keren HaYesod, 1939), 246 (Hebrew).

6. Robert R. Nathan, Oscar Gass, and Daniel Creamer, *Palestine: Problem and Promise—An Economic Study* (Washington, DC: Public Affairs Press, 1946); Michael Michaely, *Foreign Trade and Capital Imports in Israel* (Tel Aviv: Am Oved, 1963), 1–21 (Hebrew).

7. Shapira, *Futile Struggle*, 47, 153; Dov Gurevich and Aharon Gerz, *Jewish Agricultural Settlement in Pakestine* (Jerusalem: The Jewish Agency, Department of Statistics, 1938), 12–13, *14, *17, *80, *84 (Hebrew).

8. Metzer, *National Capital*; Lavsky, "Businessmen"; Idem., "Jews and Capitalism"; Idem., *Before Catastrophe*, 46–140. Various scholars have dealt with the citrus industry from different points of view. For studies that concentrate on the economic aspect of the industry see: David Horowitz, *The Economy of Palestine and its Development* (Tel Aviv: The Bialik Institute, 1944) (Hebrew); Nachum T. Gross, "Citricultre until 1939: A Success Story?," in Nachum T. Gross, *Not by Spirit Alone: Studies in the Economic History of Modern Palestine and Israel* (Jerusalem: Magnes Press and Yad Izhak Ben-Zvi, 1999), 384–89 (Hebrew); Metzer and Kaplan, *The Jewish and Arab Economies*; Metzer, *The Divided Economy*. For a thorough research on the political, as well as the social and economic aspects of the industry, see Shapira, *Futile Struggle*. From a geographical-historical perspective the industry has been studied (mainly) by Yossi Ben-Artzi, Yossi Katz, and Irit Amit. Their studies are enumerated in the bibliography.

9. Recently, a special issue of the *Journal of Israeli History* was devoted to the *Historikerstreit* in Israel, a topic that is related to a much broader *Streit*—that of the *Post Zionism Debate*. For relevant studies in this special issue pertaining to the debate over the colonialist nature of Zionism see: Derek J. Penslar, "Zionism, Colonialism, and Postcolonialism," *Journal of Israeli History* 20:2–3 (2001): 84–98; Uri Ram, "Historiographical Foundations of the Historical Strife in Israel," *Journal of Israeli History* 20:2–3 (2001): 43–61; Amnon Rav-Krakotzkin, "History Textbooks and the Limits of Israeli Consciousness," *Journal of Israeli History* 20:2–3 (2001): 155–71; Anita Shapira, "The Strategies of Historical Revisionism," *Journal of Israeli History* 20:2–3 (2001): 62–76. For earlier works on this topic see: Shapira, "Politics and Collective Memory: The Debate Over the 'New Historians' in Israel," *History and Memory* 7:1 (1995): 9–40; Baruch Kimmerling, *The Struggle for the Land: A Chapter in the Sociology of the Jewish-Arab Conflict* (Jerusalem: The Hebrew University of Jerusalem, 1974) (Hebrew); Idem., *The Economic Interrelationship between the Arab and the Jewish Communities in Mandatory Palestine* (Cambridge: Center for International Studies, MIT, 1979); Idem., *Zionism and Territory; The Socio-Territorial Dimension of Zionist Politics* (Berkeley: Institute of International Studies, University of California, 1983); Idem., *Zionism and Economy* (Cambridge, MA; Schenkman, 1983); Roger Owen, "Economic

Development in Mandatory Palestine: 1918–1948," in *The Palestinian Economy: Studies in Development under Prolonged Occupation*, ed., George T. Abed (London and New York: Routledge, 1988), 13–35; Derek J. Penslar, *Zionism and Technocracy: The Engineering of Jewish Settlement in Palestine, 1870–1918* (Bloomington and Indianapolis: Indiana University Press; 1991); Barbara J. Smith, *The Roots of Separatism in Palestine: British Economic Policy, 1920–1929* (Syracuse: Syracuse University Press, 1993); Gershon Shafir, *Land, Labor, and the Origins of the Israeli-Palestinian Conflict, 1882–1914* (Berkeley and London: University of California Press, 1996); Uri Ram, "Society and Social Science: Establishment Sociology and Critical Sociology in Israel," in *Israeli Society: Critical Perspectives*, ed. Uri Ram (Tel Aviv: Breirot, 1994), 7–39 (Hebrew); Moshe Lissak, "'Critical' Sociology and 'Establishment' Sociology in the Israeli Academic Community: Ideological Struggles or Academic Discourse?," *Israel Studies* 1:1 (1996): 247–49; Michael Shalev, *Labour and the Political Economy in Israel* (Oxford: Oxford University Press, 1992); Idem., "Time for Theory: Critical Notes on Lissak and Sternhell," *Israel Studies* 1:2 (1996): 170–88; Ran Aaronsohn, "Settlement in *Eretz Israel*—A Colonialist Enterprise? 'Critical' Scholarship and Historical Geography," *Israel Studies* 1:2 (1996): 214–29; Charles S. Kamen, *Little Common Ground: Arab Agriculture and Jewish Settlement in Palestine, 1920–1948* (Pittsburgh: University of Pittsburgh Press, 1991); Zachary Lockman, *Comrades and Enemies: Arab and Jewish Workers in Palestine, 1906–1948* (Berkeley, Los Angeles, and London: University of California Press, 1996); Ilan Pappé, "Zionism as Colonialism: A Comparative Look at Mixed Colonialism in Asia and America," in *From Vision to Revision: A Hundred Years of Historiography of Zionism*, ed. Yechiam Weitz (Jerusalem: Zalman Shazar Center, Herzl Institute for the Study of Zionism, Haifa University, and the Cherrick Center for the Study of Zionism, the Yishuv and the State of Israel, the Hebrew University of Jerusalem, 1998), 345–65 (Hebrew); Metzer, *The Divided Economy*, 200–12.

10. Theodor Herzl, *The Jewish State: An Attempt at a Modern Solution of the Jewish Question* (London: Central Office of the Zionist Organization, 1936); Moshe Smilansky, "On Questions of Colonization in Palestine," *Hashiloah* 14 (1904): 291–301, 432–42 (Hebrew); Shlomo Kaplansky, "Private Initiative and Rural Settlement," *Haahdut* 5 (Summer 1914): 2–5 (Hebrew); Penslar, *Zionism and Technocracy*. For the Mandate period see: Gideon Biger, *Crown Colony or National Homeland?: British Influence upon Palestine, 1917–1930—A Geo-Historical Analysis* (Jerusalem: Yad Izhak Ben-Zvi, 1983) (Hebrew); Jacob Reuveny, *The Administration of Palestine under the British Mandate, 1920–1948: An Institutional Analysis* (Ramat Gan: Bar-Ilan Univesity Press, 1993) (Hebrew); Smith, *Roots of Separatism*.

11. See the in-depth study by Jacob Metzer (Metzer, *The Divided Economy*).

12. Nachum T. Gross, "The Economic Policy of the Mandatory Government in Palestine," *Research in Economic History* 9 (1984): 143–85.

13. Yair Aharoni, "Some Notes on Entrepreneurship and Entrepreneurs," *Economic Quarterly* 14 (1967): 274–86 (Hebrew); Peter Kilby, "Hunting the Heffalump," in *Entrepreneurship and Economic Development*, ed., Peter Kilby (New York: The Free Press, 1971), 2; David S. Landes, "Introduction: On Technology and Growth," in *Favorites of Fortune: Technology, Growth, and Economic Development since the Industrial Revolution*, ed. Patrice Higonnet, David S. Landes, Henry Rosovsky (Cambridge and

London: Harvard University Press, 1991), 1–29; Peter Temin, "Entrepreneurs and Managers," in *Favorites of Fortune: Technology, Growth, and Economic Development since the Industrial Revolution*, ed. Patrice Higonnet, David S. Landes, Henry Rosovsky (Cambridge and London: Harvard University Press, 1991), 142–58.

14. Joseph A. Schumpeter, "Economic Theory and Entrepreneurial History," in *Change and the Entrepreneur: Postulates and Patterns for Entrepreneurial History*, Prepared by the Research Center in Entrepreneurial History, Harvard University (Cambridge: Harvard University Press, 1949), 63–84; Idem., "The Instability of Capitalism," in *The Economics of Technological Change: Selected Readings*, ed. Nathan Rosenberg (Great Britain: Penguin, 1971): 13–42; Idem., "The Fundamental Phenomenon of Economic Development," in *Entrepreneurship and Economic Development*, ed. Peter Kilby (New York: The Free Press, 1971), 43–70; Richard V. Clemence and Francis S. Doody, *The Schumpeterian System* (Cambridge, MA: Addison-Wesley, 1950).

15. Fritz Redlich, "Entrepreneurship in the Initial Stages of Industrialization (with special reference to Germany)," in *Entrepreneurship and Economic Growth* (Cambridge: The Harvard University Research Center in Entrepreneurial History, 1954), Section B; Idem., "Innovation in Business: A Systematic Presentation," in Fritz Redlich, *Steeped in Two Cultures: A Selection of Essays* (New York and Evanston: Harper and Row, 1971), 163–71; Idem., "Entrepreneurial Typology," in Fritz Redlich, *Steeped in Two Cultures: A Selection of Essays* (New York and Evanston: Harper and Row, 1971), 112–32.

16. David S. Landes, *The Unbound Prometheus: Technological Change and Industrial Development in Western Europe from 1750 to the Present* (Cambridge: Cambridge University Press, 1969), 527–28; Idem., *The Wealth and Poverty of Nations: Why Some Are so Rich and Some so Poor* (London: Little, Brown, 1998).

17. Arcadius Kahn, "Notes on Jewish Entrepreneurship in Tsarist Russia," in *Entrepreneurship in Imperial Russia and the Soviet Union*, ed. Gregory Guroff and Fred Carstensen (Princeton: Princeton University Press, 1983), 104–24.

18. Temin, "Entrepreneurs and Managers"; William Lazonick, "What Happened to the Theory of Economic Development?," in *Favorites of Fortune: Technology, Growth, and Economic Development since the Industrial Revolution*, ed. Patrice Higonnet, David S. Landes, Henry Rosovsky (Cambridge and London: Harvard University Press, 1991), 142–58.

19. Aharoni, "Some Notes on Entrepreneurship"; Alexander Gerschenkron, *Economic Backwardness in Historical Perspective: A Book of Essays* (Cambridge: Harvard University Press, 1962), 52–71; Paul H. Wilken, *Entrepreneurship: A Comparative and Historical Study* (Norwood, NJ: Ablex, 1979).

20. Nachum T. Gross, "Entrepreneurship of Religious and Ethnic Minorities," in *Juedische Unternehmer in Deutschland im 19. und 20. Jahrhundert*, ed. Werner E. Mosse and Hans Pohl (Stuttgart: F. Steiner, 1992), 11–23; Edna Bonacich, "A Theory of Ethnic Antagonism: The Split Labor Market," *American Sociological Review* 37 (1972): 547–59: Idem., "The Other Side of Ethnic Entrepreneurship: A Dialogue with Waldinger, Aldrich, Ward, and Associates," *International Migration Review* 7 (1993):

685–92; Edna Bonacich and John Modell, *The Economic Basis of Ethnic Solidarity: Small Business in the Japanese American Community* (Berkeley: University of California Press, 1980); Roger Waldinger, "The Two Sides of Ethnic Entrepreneurship," *International Migration Review* 7 (1993): 692–701; Roger Waldinger, Howard Aldrich, Robin Ward et al., *Ethnic Entrepreneurs: Immigrant Business in Industrial Societies* (Newbury Park, CA: SAGE Publications, 1990); Michael Graetz, *The Jews in Nineteenth-Century France: From the French Revolution to the Alliance isráelite universelle* (Stanford: Stanford University Press, 1996); Wilken, *Entrepreneurship*.

21. Wilken, *Entrepreneurship*, 8–23.

22. Temin, "Entrepreneurs and Managers"; Jonathan Hughes, "Public Sector Entrepreneurship," in *Favorites of Fortune: Technology, Growth, and Economic Development since the Industrial Revolution*, ed. Patrice Higonnet, David S. Landes and Henry Rosovsky (Cambridge and London: Harvard University Press, 1991), 297–321; Alexander Eckstein, "Introduction," in *Comparison of Economic Systems; Theoretical and Methodological Approaches*, ed. Alexander Eckstein (Berkeley: University of California Press, 1971), 1–23.

23. Gurevich and Gerz, *Jewish Agricultural Settlement*, 179, *84; William N. Hazen, *The Citrus Industry of Palestine* (Washington, DC: United States Department of Agriculture, Bureau of Agricultural Economics, December 1938), 1, 36; Harry Viteles, "Expansion of the Orange Industry in Palestine," *Bulletin of the Palestine Economic Society* 3:1 (1928): 3–104.

24. Hazen, *The Citrus Industry of Palestine*, 8; Shmuel Yedidyah, *Citrus Growing* (Tel Aviv: Hasadeh, 1937) (Hebrew), 30–42, 57–82.

CHAPTER TWO. DEGANIA OR PETAH TIVKA?

1. Alexis de Tocqueville, *Democracy in America*, trans. George Lawrence (New York: Harper and Row, 1966).

2. Horowitz and Lissak, *Origins of the Israeli Polity*; Farmers Federation, *Annual Reports of the Jewish Farmers Federation in Palestine* (Tel Aviv: Jewish Farmers Federation, 1931–1940) (Hebrew); CZA, A32/89.

3. Shapira, *Futile Struggle*, 310–12; Yehuda Slutsky, *The History of the "Hagana" Organization*, vol. 2 (Tel Aviv: Hassifriya Haziyonit and Ma'arachot, 1965) (Hebrew), 722–34.

4. See, for example, Shapira, *Futile Struggle*, 59–64. On the issue of "Hebrew Labor" see: Israel Kolatt, *Ideology and the Impact of Realities upon the Jewish Labour Movement in Palestine, 1905–1919* (Unpublished PhD dissertation, The Hebrew University of Jerusalem, 1964) (Hebrew); Idem., "Ha-po'el ha-Tsa'ir: From Conquest of Labor to Sanctification of Labor," *Ba-Derekh* 1 (1967): 29–61 (Hebrew); Yosef Gorny, "The Ideology of the 'Conquest of Labor,'" *Keshet* 10:2 (1968): 66–79 (Hebrew); Giladi, "The Private Farmers' Stand"; Zeev Tzahor, "Farmers and Labourers of the Second Aliya in Petah Tivka," *Cathedra* 10 (1979): 142–50 (Hebrew); Shabtai Teveth, *David's Jealousy: The Life of David Ben Gurion*, vol. 2 (Jerusalem and Tel Aviv: Schoken, 1980) (Hebrew), 422–55; Dalia Hurwitz, "The Attitude of the Zionist Insti-

tutions and Leaders in Palestine to Jewish Labour during the Period of the Second Aliya (1904–1914)," *Zionism* 7 (1981): 95–134 (Hebrew); Shafir, *Land, Labor*; Katz, *Zionist Private Enterprise*; Shapira, *Futile Struggle*. It should be emphasized that it is not my purpose in this study to delve once again into this topic of "Hebrew Labor." It will be touched upon, however, when necessary whenever it be relevant to the main argument of this book.

5. On Smilansky's political views and career see: Shapira, *Futile Struggle*; Esther Reisen, "High Hopes: The Beginning of Moshe Smilansky's Public Career" (Unpublished MA thesis, Tel Aviv, Tel Aviv University, 1982) (Hebrew); Yossef Gorny, *Zionism and the Arabs, 1882–1948: A Study of Ideology* (Oxford: Clarendon Press, 1987); Neal Sherman and Daniella Heller, "The Establishment and Institution of the Farmers' Federation of Eretz Israel," *Cathedra* 58 (1990): 155–71 (Hebrew). On his literary works see: Rissa Dom, "The Character of the Arab in Moshe Smilansky's Literary Works," *Iton 77* 84–85 (1987): 95–97 (Hebrew); Yaffah Berlovitz, "Moshe Smilansky—'Benei Arav: The Patronizing Stories," *Iyunim Bitkumat Israel* 4 (1994): 400–21 (Hebrew); Idem., *Inventing a Land, Inventing a People* (Tel Aviv: Hakibbutz Hameuchad, 1996) (Hebrew), 148–75; Ido Bassok, *In the Shadow of Orchards on a Leprous Soil: Studies in the Literary and Essayitic work of Moshe Smilansky and Uri Zvi Grinberg* (Tel Aviv: Hakibbutz Hameuchad, 1996) (Hebrew); Pinhas Ginossar, "The Community of Hebrew Writers in Eretz-Israel, 1918–1923," *Iyunim Bitkumat Israel* 6 (1996): 479–534 (Hebrew). On his biography, including his autobiographical novels see: S. Yizhar, *Mikdamot* (Tel Aviv: Zmora-Bitan, 1992) (Hebrew), 192–98; Interviews with his daughter Rina Smilansky and nephew, the Israeli writer S. (Smilansky) Yizhar; David Smilansky, *A City Is Born: The Story of Tel Aviv and Eretz-Israel in the Second Aliya Period, According to the Letters of David Smilansky as Published in the Press in Russia*, ed. and introduction Yossi Katz (Tel Aviv: Ministry of Defense Publishing House, 1981) (Hebrew); Moshe Smilansky, *In the Fields of Ukraine: A Story in Five Parts* (Tel Aviv: Massada, 1944) (Hebrew); Idem., *In the Wilderness: A Story* (Tel Aviv: Massada, 1947 ?) (Hebrew); Idem., *In the Shadow of the Orchards: A Story* (Tel Aviv: Massada, 1951 [?]) (Hebrew); Idem., *Resurrection and Holocaust: A Story* (Tel Aviv: Massada, 1953) (Hebrew); Idem., *Among the Vineyards of Judea: A Story* (Tel Aviv: Massada, 1954) (Hebrew); Idem., *Birth Pangs: A Story* (Tel Aviv: Massada, 1954) (Hebrew); Idem., "Personal Questionnaire," *Genazim* 7 (April 1979): 276–79 (Hebrew).

6. Smilansky, "On Questions of Colonization in Palestine" *Hashiloah* (Idem., *Writings of Moshe Smilansky*, vol. 11 [Tel Aviv: Farmers' Federation, 1937], 83–112 [Hebrew]). Quote from *Writings of Moshe Smilansky*, vol. 11, 94–96; Paul Leroy-Beaulieu, *De la Colonisation chez les peuples modernes* (Paris: Guillaumin, 1902).

7. *Writings of Moshe Smilansky*, vol. 11, 83–112; Berlovitz, "Beni Arav"; Moshe Smilansky, "Jewish Colonization and the Fellah," *Palestine and Near East Economic Magazine* 5 (1930).

8. For a similar—and much more famous—approach to the Chartered Company as a colonizing tool in Zionist settlement see: Herzl, *The Jewish State*.

9. Smilansky, as well as other Zionist leaders, was familiar with the German colonizing project in Poznan. The formal control of these former Polish lands did not,

ipso facto, allow for a full-fledged German settlement of the Poznan province. Thus, the Prussian and the German governments set out to settle these lands by Germans through a comprehensive project of "Inner Colonization." On the Poznan colonizing project as a model for Zionist settlement see: Shalom Reichman and Shlomo Hasson, "A Cross-Cultural Diffusion of Colonization: From Posen to Palestine," *Annals of the Association of American Geographers* 74:1 (March 1984): 57–70; Penslar, *Zionism and Technocracy*, 94–98; Shafir, *Land*, 146–60; Zvi Shilony, "German Antecedents of Rural Settlement in Palestine up to World War I," in *The Land that Became Israel; Studies in Historical Geography*, ed. Ruth Kark (New Haven, London, and Jerusalem: Yale University Press and The Magnes Press, 1989), 196–214.

10. Margalit Shilo, *Experiments in Settlement* (Jerusalem: Yad Izhak Ben-Zvi, 1988) (Hebrew), 12–15; Zalman David Levontin, *To the Land of Our Fathers*, 3 vols. (Tel Aviv: Eitan and Shohsany, 1924–1928) (Hebrew), vol. 2, 93; Gross, "Private and Public Enterprise," 488–94.

11. Kolatt, *Ideology*; Yigal Drori, "The Second Aliya and the Struggle for Jewish Labour as Reflected in the General Eretz Israel Press," *Cathedra* 2 (1977): 69–80 (Hebrew). Shapira, *Berl Katznelson*, vol. 1, 45–117; Idem. *Futile Struggle*, 15–32.

12. David Ben Gurion, *From Class to Nation: Reflections on the Goals and Policies of the Workers' Movement* (Tel Aviv: Davar, 1933) (Hebrew), 27–28.

13. Arthur Ruppin, *Thirty Years of Building in Palestine* (Jerusalem: Schocken, 1937) (Hebrew), 1–62; Shilo, *Experiments in Settlement*; Penslar, *Zionism and Technocracy*, 80–154.

14. Ruppin, *Thirty Years*, 39–47; Metzer, *National Capital*, 23; Penslar, *Zionism and Technocracy*, 90–91.

15. Ahad Ha-'Am, (Asher Ginsberg), "All in All," *Hashiloah* 26 (1914): 276–90.

16. Ibid.; the quote from: Idem., *Collected Writings of Ahad Ha-'Am (Kol kitvei Ahad Ha-'Am)* (Tel Aviv: Dvir, 1957) (Hebrew), 425. On the concept of "National Existence Spirit" in Ahad Ha-'Am's thought and on his philosophical outlook and leadership position see: Yehiel Alfred Gottschalk, *Ahad Ha-'Am and the National Spirit* (Jeruslaem: Hassifriya HaZiyonit, 1992) (Hebrew); Steven Jeffrey Zipperstein, *Elusive Prophet: Ahad Ha'am and Origins of Zionism* (Berkeley: University California Press, 1993).

17. Ahad Ha-'Am, "All in All," 425.

18. Ahad Ha-'Am, "Truth from Eretz-Israel—Part II," in *Collected Writings*, 31.

19. Yizhak Wilkansky, "On the Expectations for Capital," *Hapoel Hatzair* 6:35 (1913): 3–6; 6:36 (1913): 5–9 (Hebrew); On Wilkansky see: Penslar, *Zionism and Technocracy*, 123–49.

20. Moshe Smilansky, "With the Flow of Phrase," *Ha-Herut* 5:293–97 (1913), in segments (Hebrew).

21. Smilansky, *Collected Writings*, vol. 12, 103–59.

22. On European colonialism at that time see: David Kenneth Fieldhouse, *The Colonial Empires: A Comparative Survey from the Eighteenth Century* (London: Wei-

denfeld and Nicolson; 1966); Idem., *Colonialism 1870–1945: An Introduction* (London: Weidenfeld and Nicolson, 1981).

23. Moshe Smilansky, "Our Capitalistic Economy," in *Collected Writings*, vol. 12, 58–59.

24. Smilansky, "Our Capitalistic Economy"; Idem., "Real Activities or Conditions for Future Activities?," in *Collected Writings*, vol. 12, 75–102.

25. Smith, *The Roots of Separatism.*

26. Metzer, *National Capital*, 32–39, 141–43; Shapira, *Berl*, vol. I, 172–77; Hagit Lavsky, "Theory and Praxis: The Agrarian Policy of the Jewish National Fund during the Mandatory Period," in *Studies in Geography and History in Honour of Yehoshua Ben-Arieh*, ed. Yossi Ben-Arzi, Israel Bartal, and Elchanan Reiner (Jerusalem: Magnes Press and Israel Exploration Society, 1999) (Hebrew), 438–60.

27. Ben Gurion, *From Class to Nation*, 17.

28. The research literature on the economic-socialist outlook of the Labor Movement is extensive, see: Arthur Ruppin, *Agricultural Settlement of the Zionist Organization in Palestine (1918–1924)* (Tel Aviv: Dvir, 1925) (Hebrew), 1–9; Yosef Gorny, *Achdut Haavoda, 1919–1930: The Ideological Principles and the Political System* (Tel Aviv: Hakibbutz Hameuchad, 1973) (Hebrew), 11–113; Haim Golan (ed.), *The Poale-Zion Commission in Palestine—1920* (Ramat Efal: Yad Tabenkin, 1989) (Hebrew); Bein, *History of Zionist Settlement*, 140–260; Kolatt, *Ideology*; Idem., "Ha-po'el ha-Tsa'ir," 29–61; Idem., *Fathers and Founders* (Tel Aviv: Institute for Contemporary Jewry, The Hebrew University of Jerusalem, and Hakibbutz Hameuchad, 1976) (Hebrew), 32–36; Shapira, *Futile Struggle*; Idem., *Visions in Conflict* (Tel Aviv: Am Oved, 1989) (Hebrew), 355–74; Yaacov Shavit and Dan Giladi, "The Role of the Dairy Farm in the Development of Jewish Settlement during the Mandatory Period," *Cathedra* 18 (January 1981): 178–92 (Hebrew); Sternhell, *Nation-Building or a New Society?*; Baruch Ben-Avram, "The Emergence of the Kvutza from the Aspiration of the Pioneers of the Second Aliya for Self-Employment," *Cathedra* 18 (1981): 118–23 (Hebrew); Idem. and Henry Near, *Studies in the Third Aliyah: Image and Reality* (Jerusalem: Yad Izhak Ben-Zvi, 1995) (Hebrew), 5–101; Henry Near, "Ideology, Anti-Ideology, and Historiography," *Cathedra* 18 (1981): 124–29 (Hebrew); Berl Katznelson, *Collected Writings* (Tel Aviv: Mapai, 1946–1950) (Hebrew), vol. 3, 50.

29. Nachum T. Gross and Yitzhak Greenberg, *Bank Ha'poalim: The First 50 Years, 1921–1971* (Tel Aviv: Am Oved, 1994) (Hebrew); Yitzhak Greenberg, *From Workers' Society to Workers' Economy: Evolution of the Hevrat Ha'ovdim Idea 1920–1929* (Tel Aviv: Papyrus, 1987) (Hebrew); Giladi, *Jewish Palestine during the Fourth Aliya Period*, 208–14.

30. Gorny, *Achdut Haavoda*, 71–74; Golan (ed.), *The Poale-Zion Commission in Palestine*; Sternhell, *Nation-Building or a New Society?*; Giladi, *Jewish Palestine during the Fourth Aliya Period*, 85–93; Amit, *Jewish Settlement Activity in the Southern Sharon*, 101; CZA S15/276a, S15/485.

31. Katznelson, *Collected Writings*, vol. 6, 11–39, quote from 21. See also: Ibid., vol. 4, 195–221. On this change of attitude of the labor movement, which needs a fur-

ther research, see: Ruppin, *Agricultural Settlement*; Idem., *Thirty Years*, 215–48; Bezalel Amikam, *First Tillings: The Jewish Agricultural Settlements in the 1930s* (Jerusalem: Yad Izhak Ben-Zvi, 1980) (Hebrew), 114–45; *Department of Agricultural Settlement—Zionist Organization: Reports for the 14th,15th, and 16th Zionist Congresses* (Jerusalem and London: Department of Agricultural Settlement, 1925–1929) (Hebrew).

32. Ruppin, *Agricultural Settlement*; Idem., *Thirty Years*, 238–48; Metzer, *National Capital*, 32–39; Amikam, *First Tillings*, 129–45.

33. David Ben Gurion, "Towards the Convention," *Kuntres* 14: 279 (1926): 7–8 (Hebrew). See also: Gorny, *Achdut Haavoda*, 43–93; Greenberg, *From Workers' Society*, 261–87; Gross and Greenberg, *Bank Hapoalim*, 13–86.

34. Katznelson, *Collected Writings*, vol. 3, 50.

35. Ben Gurion, *From Class to Nation*, 296.

36. Ibid.

37. Ibid., 231–33, 296–97.

38. M. Eldad, "'National' Colonization and 'Private' Colonization," *Bustenai* 4:42 (1932/33): 5; "Chaim Weizmann Speech before the Annual Convention of British Zionists," *Haolam* 21:2 (12.1.1933), 25–26; Moshe Smilansky, "A Lip Service to a 'New Word,'" *Bustenai* 4:39 (1932/33): 5–7 (Hebrew); Idem., "'Boazism' and 'National' Settlement," *Bustenai* 4:40 (1932/33): 3–5. All the following quotations are from this last article.

39. Moshe Smilansky, "A 'New Word'," *Bustenai* 5:39 (1933/34): 6.

40. Gross, "Free Enterprise Versus Planned Markets," 355–65 (Hebrew); Simon Kuznets, *Modern Economic Growth: Rate, Structure, and Spread* (New Haven and London: Yale University Press; 1966); Landes, *The Unbound Prometheus*; Idem.,"Introduction: On Technology and Growth"; Idem., *The Wealth and Poverty of Nations*; Haim Barkai, *Adam Smith in the Context of Our Times* (Jerusalem and Tel Aviv: The Bialik Institute and the Open University, Israel, 1995) (Hebrew); Nathan Rosenberg and L. E. Birdzell, *How the West Grew Rich: The Economic Transformation of the Industrial World* (New York: Basic Books, 1986).

CHAPTER THREE. SPATIAL DISTRIBUTION AND SOCIAL AND ENTREPRENEURIAL PROFILE

1. Shmuel Tolkowsky, *Citrus Fruits: Their Origin and History throughout the World* (Jerusalem: The Bialik Institute, 1966) (Hebrew), 249–52. On the definition of the term "Spatial Distribution" see: Yehoshua Ben-Aryeh, "Geographic Aspects of the Development of the First Jewish Settlements in Palestine," in *The First Aliyah*, ed. Mordechai Eliav (Jerusalem: Yad Izhak Ben-Zvi and Israel Ministry of Defense Press, 1981), 85–96 (Hebrew).

2. Shmuel Avitsur, *The Rise and Fall of the Port of Jaffa* (Tel Aviv: Milo, 1972), 28–47 (Hebrew); Nachum T. Gross, "Laying the Foundations," in *Banker for a Renewed Nation: The History of Bank Leumi Le'Israel*, ed. Nadav Halevi (Ramat Gan: Massada, 1977), 44–45 (Hebrew); Idem., "Economic Changes in Eretz Israel at the

End of the Ottoman Period," *Cathedra* 2 (1977): 111–25 (Hebrew); Idem., "The Palestine Economy in the Final Decades of Ottoman Rule," in *Not by Spirit Alone: Studies in the Economic History of Modern Palestine and Israel*, ed. Nachum T. Gross (Jerusalem: Magnes Press and Yad Izhak Ben-Zvi, 1999), 19–52 (Hebrew); Avraham Fellman, *The Pioneers of Jewish Citriculture in Palestine (the Fellman Family)* (Tel Aviv: Fellman, 1940) (Hebrew); Aharon Leib Fellman, *Ma'ayan Ganim: Rules for Planting Gardens and Citrus Orchards in the Holy Land* (Jerusalem: 1891), 8 (Hebrew).

3. Johann Heinrich Von Thünen, *The Isolated State*, trans. Carla M. Wartenberg, ed. Peter Hall (Oxford: Pergamon, 1966); John R. Tarrant, *Agricultural Geography* (New York: John Wiley and Sons, 1974); Yehuda Loew, *Agricultural Economy* (Tel Aviv: Am Oved, 1957), 43–60 (Hebrew).

4. CZA J15/6122

5. Curt Nawratzki, *Die Juedische Kolonisation Palaestinas* (Muenchen: Reinhardt, 1911), 180, 338; Shaul Katz, *Sociological Aspects of the Growth and the Turnover of Agricultural Knowledge in Israel: The Emergence of Extra-Scientific Systems for the Production of Agricultural Knowledge 1880–1940* (Unpublished PhD dissertation, the Hebrew University of Jerusalem, Jerusalem, 1986), 100–103 (Hebrew).

6. Gross, "Laying the Foundations," 41–60; Idem., "Economic Changes"; Idem., "The Palestine Economy."

7. Nawratzki, *Juedische Kolonisation*, 122–201; David Yudilowitz (ed.), *Rishon Lezion, 1882–1941* (Rishon Lezion: Carmel Mizrahi, 1941) (Hebrew); Ran Aaronsohn, *Rothschild and Early Jewish Colonization in Palestine* (Jerusalem: Magnes Press, 2000); Levontin, *To the Land of Our Fathers*, vol. I, 15–61 (Hebrew); Ruth Kark, "Transportation in Nineteenth-Century Palestine: Reintroduction of the Wheel," in *The Land that Became Israel: Studies in Historical Geography*, ed. Ruth Kark (New Haven, London and Jerusalem: Yale University Press and The Magnes Press, 1989), 57–76.

8. Gurevich and Gerz, *Jewish Agricultural Settlement*, *34, *43–44.

9. On Shimon Rokach see: Isaac Rokach, *Tales of the Jaffa Orange Groves* (Ramat Gan: Massada, 1970), especially 101–14 (Hebrew). Ruth Kark, *Jaffa: A City in Evolution, 1799–1917* (Jerusalem: Yad Izhak Ben-Zvi, 1990); Shulamit Laskov, "The 'Moshavot' under the Tutelage of 'Hovve Zion' and Those Who Were Not," in *The History of the Jewish Community in Eretz-Israel since 1882: The Ottoman Period*, ed. Israel Kolatt, part one (Jerusalem: The Israel Academy for Sciences and Humanities, The Bialik Institute, 1989), 393–407 (Hebrew); Yoseph Salmon, "The Urban Ashkenazi Community in Palestine, 1880–1903," in *The History of the Jewish Community in Eretz-Israel since 1881: The Ottoman Period*, ed. Israel Kolatt, part one (Jerusalem: The Israel Academy for Sciences and Humanities, The Bialik Institute, 1989), 554–619 (Hebrew); CZA A323/16.

10. On Mazie see: Aaronsohn, *Rothschild and Early Jewish Colonization in Palestine*; Idem., "Baron Rothschild's Officials, 1882–1890," *Cathedra* 74 (1994): 157–78 (Hebrew); Nachum Slushtz, *Dr. Aharon Meir Mazie: His Life and Activities* (Jerusalem: Jewish Exploration Society of Israel, 1935) (Hebrew); Eliahu Izakson, *A View from the Bridge Generation* (Tel Aviv: Eliahu Izakson, 1994) (Hebrew); Aharon Fellman, *Fifty*

Years of the Pardess Society, Fifty Years of Jewish Citriculture: 1900–1950 (Tel Aviv: Pardess Society, 1950), 92–95 (Hebrew).

11. CZA J15/6129, J15/6122; Katz, *Sociological Aspects*, vol. II, 100–103; Rokach, *Tales*, 95–116; Fellman, *Fifty Years of the Pardess Society*, 32; Getzel Kressel, *Petah Tikva, Mother of Settlements, 1878–1953* (Tel Aviv: Municipality of Petah Tikva, 1953) (Hebrew).

12. Rehovot Municipal Archives; interview with Deborah Dimant, daughter of Tovia Miller; Moshe Smilansky, *Rehovot: Its Sixty Years of Life, 1890–1950* (Rehovot: Rehovot Local Council, 1950) (Hebrew).

13. Interview with Ada Aurebach and Dan Tolkowsky, Shmuel Tolkowsky's children; Smilansky, *Rehovot*, 54–68; Amit, *The Southern Sharon*, 290–313; CZA A248/5, 12.

14. The Holtzmann archives; interview with Holtzmann's daughter, Ahuvah Zmora; Zeev Aner, "Holzmann: The Man Who Gave His Name to the Etzion Bloc," in *The Etzion Bloc from Its Inception to 1948 (Idan 7)*, ed. Mordechai Naor (Jerusalem: 1986), 53–62 (Hebrew); Zvi Ilan, "Holzmann's Partners in Establishing Kefar 'Etzyon in the 1930s," in *The Etzion Bloc from Its Inception to 1948 (Idan 7)*, ed. Mordechai Naor (Jerusalem: 1986), 63–74 (Hebrew); Margalit Shilo, "Dr. Arthur Ruppin and Settlement in the Mountain Area: His Attitude toward the 'To the Mountains' Affair and the First Kefar Etzion," in *The Etzion Bloc from Its Inception to 1948 (Idan 7)*, ed. Mordechai Naor (Jerusalem: 1986), 41–52 (Hebrew).

15. On the Jacobsohn family see: CZA A249; private archives of Zalman Jacobsohn; Rokach, *Tales*, 213–16; interviews with Jacobsohn's children—Hill Jacobsohn, Imanuel Jacobsohn, Daniel Jacobsohn, Leah Zuker.

16. Katz, *Sociological Aspects*, vol. I, 149–50, 319.

17. Gurevich and Gerz, *Jewish Agricultural Settlement*, *32, *34; D. Wolochonsky,, "Rishon Lezion in the Lights of the Statistics," *Mishar Ve-Taasiyah* 8:4–5 (1930): 39–62 (Hebrew).

18. CZA A238/12.

19. CZA L18/228/2.

20. Avitsur, *The Port of Jaffa*, 136–46; Idem., *The Industrial Revolution in Eretz Israel in Its Energetic Processes* (Tel Aviv: Hakibbutz Hameuchad, 1989), 206–44 (Hebrew); Gross, "The Economic Policy of the Mandatory Government, 143–85; Shalom Reichman, "The Evolution of Land Transportation in Palestine, 1920–1947," *Jerusalem Studies in Geography* 2 (1971): 55–90; Biger, *Crown Colony or National Homeland?*, 96–113, 185–87 (Hebrew); Fruit Commission, *Report of the Fruit Commission* (Palestine: Government of Palestine, 1927).

21. Gurevich and Gerz, *Jewish Agricultural Settlement*, 61; Wolochonsky, "Rishon Lezion"; Smilansky, *Rehovot*, 63 onward; Table 5.2 above; Farmers Federation, *Annual Reports*; Amiram Gonen, "Urbanization of the 'Moshavoth' on Israel's Coastal Plain: Factors and Stages," in *Studies in Geography and History in Honour of Yehoshua Ben-Arieh*, ed. Yossi Ben-Arzi, Israel Bartal, and Elchanan Reiner (Jerusalem: Magness Press and Israel Exploration Society, 1999), 461–77 (Hebrew).

22. Yakhin Cooperative Society, *Statutes* (Tel Aviv: Yakhin, 1929) (Hebrew); Idem., *Four Years of Activities of the Yakhin Cooperative Society for Agricultural Cultivation, 1927–1930* (Tel Aviv: Yakhin, 1931) (Hebrew).

23. Yakhin Cooperative Society, *Annual Report for 1930*, 1.

24. Labor Archives, IV–208–1–1070; Yakhin Cooperative Society, *Annual Reports for the years 1933–38*.

25. *Bustenai*, 2:28 (1930/1): 21.

26. CZA S15/276a, S15/485; Yigal Drori, *The "Civil Circles" in the Yishuv during the 1920s* (Tel Aviv: Mif'alim Universitaim, 1990), 112–79 (Hebrew); *Bustenai* 2:30 (1930/1): 6–7; Ibid., 4:38 (1933/4): 21; Ibid., 9:8 (1937/8): 28–33; Hanotaiah, *Netanyah and What It Symbolizes* (Tel Aviv: Hanotaiah, 1932) (Hebrew); Idem., *Orange Groves in Palestine* (Tel Aviv: Hanotaiah, 1933); Rahel Kleinman, Yehoshua Ben-Aryeh, and Dan Giladi, "The Beginnings of Netanya as an Agricultural Colony, 1928–1933," in *Netanya Book*, ed. Avshalom Shmueli and Moshe Brawer (Tel Aviv: Am Oved, 1982), 115–22 (Hebrew); Idem., "Initial Development of an Urban Settlement alongside the Colony of Netanya, 1933–1939," in Avshalom Shmueli and Moshe Brawer (eds.), *Netanya Book* (Tel Aviv: Am Oved, 1982), 123–38 (Hebrew); Yossi Ben-Artzi and Abrahahm Labes, "Netanyah's First Plan," *Ofakim Be-Geographiyah* 6 (1982): 65–68 (Hebrew).

27. Ben-Yisrael (Ze'ev Smilansky), "Citrus Groves Then and Now," *Bustenai* 3:45 (1931/2): 16; Gurevich and Gerz, *Jewish Agricultural Settlement*, 102.

28. Ibid.

CHAPTER FOUR. FROM JAFFA TO PETAH TIKVA

1. Landes., *The Unbound Prometheus*; Idem., "Introduction: On Technology and Growth," 1–29; Idem., *The Wealth and Poverty of Nations*; Joel Mokyr, *The Lever of Riches: Technological Creativity and Economic Progress* (New York and Oxford: Oxford University Press, 1990); Idem., "Technological Inertia in Economic History," *Journal of Economic History* 52 (1992): 325–38; Idem., *The Gifts of Athena: Historical Origins of the Knowledge Economy* (Princeton: Princeton University Press, 2002); Nathan Rosenberg, *Technology and American Economic Growth* (New York: Harper and Row, 1972); Idem., *Inside the Black Box: Technology and Economics* (Cambridge: Cambridge University Press; 1982); Idem., *Exploring the Black Box; Technology, Economics, and History* (Cambridge: Cambridge University Press; 1994).

2. Theodore W. Schultz, *Transforming Traditional Agriculture* (New Haven and London: Yale University Press, 1969); Yujiro Hayami and Vernon W. Ruttan, *Agricultural Development: An International Perspective* (Baltimore and London: Johns Hopkins University Press, 1971); Raanan Weitz, *From Peasant to Farmer: A Revolutionary Strategy for Development* (New York and London: Columbia University Press, 1971); Michael Todaro, *Economic Development* (New York and London: Longman, 1994).

3. Metzer, *National Capital*; Idem., "Technology, Labor, and Growth in a Dual Economy's Traditional Sector: Mandatory Palestine, 1921–1936," in *Technical Change*,

Employment, and Investment, ed. Lennart Joerberg and Nathan Rosenberg (Lund: Lund University, 1982), 159–70; Idem., "What Kind of Growth? A Comparative Look at the Arab Economies in Mandatory Palestine and in the Administered Territories," *Economic Development and Cultural Change* 40 (1992): 843–65; Idem., *The Divided Economy*; Metzer and Kaplan, *The Jewish and Arab Economies*; Beenstock, Metzer, and Ziv, "Immigration and Jewish Economy," 149–213; Katz, *Sociological Aspects*; Gur Ofer, *The Service Industries in a Developing Economy: Israel as a Case Study* (New York; F. A. Praeger, 1967); Zvi Sussman, *Wage Differentials and the Equality within the Histadrut: The Impact of Egalitarian Ideology and Arab Labor on Jewish Wages in Palestine* (Ramat Gan: Massada, 1974) (Hebrew).

4. Rosenberg, *Technology*; Idem., *Inside the Black Box*; Idem., *Exploring the Black Box*.

5. See Shmuel Avitsur, Amiram Oren, Ran Aaronsohn, Simon Schama, Naftali Thalmann and Shaul Katz studies, listed in the bibliography and, Ruth Kark, "The Introduction of Modern Technology into the Holy Land (1800–1914)," in *The Archaeology of Society in the Holy Land*, ed. Thomas E. Levy (London: Leicester University Press, 1995), 524–41.

6. Mokyr, *The Lever of Riches*, 3–16; Rosenbeg, *Inside the Black Box*; Vernon Ruttan, "Usher and Schumpeter on Invention, Innovation and Technological Change," in *The Economics of Technological Change: Selected Readings*, ed. Nathan Rosenberg (Great Britain: Penguin, 1971), 73–85; Idem., *Technology, Growth, and Development: An Induced Innovation Perspective* (New York and Oxford: Oxford University Press, 2001), esp., 61–234; Schumpeter, "Economic Theory and Entrepreneurial History"; Idem., "The Instability of Capitalism."

7. Mokyr, *The Lever of Riches*, 3–16.

8. Colin G. Thirtle, and Vernon W. Ruttan, *The Role of Demand and Supply in the Generation and Diffusion of Technical Change* (Chur: Hardwood Academic Publishers, 1987); Landes, *The Unbound Prometheus*; Idem., "Introduction"; Idem., *The Wealth and Poverty of Nations*; Schultz, *Transforming Traditional Agriculture*; Zvi Griliches, "Hybrid Corn and the Economics of Innovation," in *The Economics of Technological Change: Selected Readings*, ed. Nathan Rosenberg (Great Britain: Penguin, 1971), 211–28; Rosenberg, *Technology*; Idem., *Inside the Black Box*; Idem., *Exploring the Black Box*; Kuznets, *Modern Economic Growth*; Mokyr, *The Lever of Riches*; Idem., *The Gifts of Athena*.

9. All quotes and references are from the following: Fellman, *Ma'ayan Ganim*; G. Franghia, "Report on Irrigation and Orange Growing at Jaffa," Great Britain: Foreign Office, *Miscellaneous Series no. 300, Reports on Subjects of General and Commercial Interest* (London: 1893); Aaron Aaronsohn and Selig Soskin, "Die Orangengaerten von Jaffa," *Der Tropenpflanzer* 4 (1902): 341–61; Shmuel Avitsur, "The First Project for the Intensive Exploitation of the Yarkon Waters (The Frangija-Navon Scheme of 1893)," *Museum Haaretz Bulletin* 6 (1964): 80–88. See also: Mordechai Diakin, *The Words of Mordechai* (Jerusalem: 1889) (Hebrew).

10. For example, Aaronsohn and Soskin, "Jaffa," 355; Smilansky, "Jewish Colonization and the Fellah"; see critique of this approach in Schultz, Weitz, and Todaro,

cited in the bibliography; Schultz, *Transforming Traditional Agriculture*; Rosenberg, *Exploring the Black Box*.

11. Fellman, *Ma'ayan Ganim*; Aaronsohn and Soskin, "Jaffa"; Franghia, "Report."

12. Shmuel Avitsur, *Changes in Agriculture of Eretz Israel 1875–1975* (Tel Aviv: Milo,

1978) (Hebrew); Idem., "The Contribution of Early Petah Tiqva to the Agricultural and Industrial Development of Eretz Israel (1878–1918)," *Cathedra* 10 (1979): 129–41 (Hebrew); Idem., "A Pioneer of Modern Industry in Eretz Israel: The History of the L. Stein Factory," *Cathedra* 15 (1980): 69–94 (Hebrew); Peretz Pascal, "Citriculture in the Jubilee of Petah Tikva," *Bustenai* 10: 27 (1938): 31–32 (Hebrew); Fellman, *Ma'ayan Ganim*, 8–17, 24; Aaronsohn and Soskin, "Jaffa," 350–51.

13. Fellman, *Ma'ayan Ganim*, 9–13.

14. Ibid.; Aaronsohn and Soskin, "Jaffa," 352.

15. Fellman, *Ma'ayan Ganim*, 11–31.

16. Ibid.,; Aaronsohn and Soskin, "Jaffa," 351; Franghia, "Report," 3–4.

17. Aaronsohn and Soskin, "Jaffa"; Naftali Thalmann, "Introducing Modern Agriculture into Nineteenth-Century Palestine: The German Templers," in *The Land that Became Israel: Studies in Historical Geography*, ed. Ruth Kark (New Haven, London, and Jerusalem: Yale University Press and The Magnes Press, 1989), 90–104.

18. Naftali Thalmann, *The Character and Development of the Farm Economy in the Templer Colonies in Palestine, 1869–1939* (Unpublished PhD dissertation, Hebrew University of Jerusalem, Jeruslaem, 1991), 89 (Hebrew); Aaronsohn, *Rothschild and Early Jewish Colonization*.

19. Israel State Archives, Bloc 7 (Department of Agriculture), container *mem*-675, file H/64/21/8; Rokach, *Tales*, 37 (Hebrew).

20. Gross, "Economic Changes; Aaron Aaronsohn, *Citrus Fruit around the Globe as It Seems from the Palestinian Grower's Perspective* (Jaffa: The Zionist Organization, Agricultural Experimental Station, 1915) (Hebrew).

21. Aaronsohn and Soskin, "Jaffa," 358; Avitsur, *Changes*, 116–17.

22. Thalmann, *Character and Development*, 152–70; Avitsur, "Stein," 76; Idem., *Changes*, 183; Nawratzki, *Die Juedische Kolonisation*, 174–75, 335; Thirtle and Ruttan, *The Role of Demand and Supply*, 77–89.

23. Rosenberg, *Inside the Black Box*, 120–40; Yoav Kislev and Willis Peterson, "Prices, Technology, and Farm Size," *Journal of Political Economy* 90 (1982): 578–95; Aaronsohn and Soskin, "Jaffa," 358–59. In regard to the *manége*, see Avitsur, *The Industrial Revolution*, 48–50, 75–78.

24. Ibid.; Aaronsohn, *Rothschild*.

25. Nawratzki, *Juedische Kolonisation*, 123, 392–93.

26. Meir Apfelbaum, "Citrus Growing in Palestine," *HaHaklai* 1 (1912/13): 218 (Hebrew); Nawratzki, *Juedische Kolonisation*, 336.

27. Apfelbaum, "Citrus." For a principled discussion of this topic, see Jacob Metzer, "Rational Management, Modern Business Practices, and Economies of Scale in the Ante-Bellum Southern Plantation," *Exploration in Economic History* 12 (1975): 123–50; Gershon Feder and Gerald T. O'Mara, "Farm Size and the Diffusion of Green Revolution Technology," *Economic Development and Cultural Change* 30: 1 (1981): 59–76.

28. Avitsur, *Changes*, 111–49; Idem., "Petah Tiqva," 130–37; Melekh Zagorodsky, *The Citrus Book: Manual for Orange Growers, Workers, Experts, and Nurseries Owners* (Tel Aviv: Hahaklai, 1929), B. 41–43 (Hebrew).

29. Apfelbaum, "Citrus," 5–6, 47, 116; Avitsur, *Industrial Revolution*, 44–48, 55–57, 74–77, 131–33; and see Davies, David, and Mansfield's studies cited in the bibliography.

30. Aaronsohn and Soskin, "Jaffa," 359; Apfelbaum, "Citrus," 6; Moshe Smilansky, "Planting an Orchard," *Hasadeh* 10 (1926): 267 (Hebrew); Katz, *Sociological Aspects*, A, ch. 7.

31. Apfelbaum, "Citrus," 6, 115–16; Shimshon Katzparovsky, "What One May Learn from California about Growing Citrus Trees in Palestine," *Ha-haklai* 5: 3 (Summer 1918): 7–12; Rokach, *Tales of the Jaffa Orange Groves*, 110–14; Zagorodsky, *Citrus Book*, B, 1–26.

32. Yedidyah, *Citrus Growing*, 355–58; Apfelbaum, "Citrus," 4–5, 116–18; Zagorodsky, *Citrus Book*, A, 67–184.

33. Apfelbaum, "Citrus," 118, 216–17; CZA, A323/435.

34. Apfelbaum, "Citrus," 218; Moshe Smilansky, "The New Yishuv," in *Writings of Moshe Smilansky*, vol. 11, 118–19.

35. For discussion of this period, see Fellman, *Fifty Years of the Pardess Society*, 62–81; Pardess Society, *Annual Reports of the Pardess Society* (various years).

CHAPTER FIVE. TECHNOLOGICAL INNOVATION DURING THE MANDATE ERA

1. Nahum Karlinsky, "California Dreaming: Adapting the 'California Model' to the Jewish Citrus Industry in Palestine, 1917–1939," *Israel Studies* 5:1 (2000): 24–40.

2. Shapira, *Futile Struggle*.

3. Interview with Devora Dimant, Tovia Miller's daughter; Shmuel Zakif, *Diaspora and Homeland* (Magdiel: Magdiel Municipality, 1954) (Hebrew); CZA, A189 (Zakif archives).

4. Metzer and Kaplan, *The Jewish and Arab Economies*, 39.

5. Moshe Smilansky, "The Last Line," *Bustenai* 9: 4 (May 5, 1937): 3–4.

6. Knowles A. Ryerson, "The Horticultural Possibilities as Especially Related to Agricultural Colonization," *Reports of the Experts: Submitted to the Joint Palestine Survey Commission* (Boston: Daniels Printing Co., 1928), 268. See also Ryerson's memoirs from his visit to Palestine: Knowles A. Ryerson, *The World is my Campus* (Uni-

versity of California Davis, Shields Library: The Oral History Center, 1977), 146–59. On the delegation of Experts, see Ilan S. Troen, "American Experts in the Design of Zionist Society: The Reports of Elwood Mead and Robert Nathan," in *Envisioning Israel: The Changing Ideas and Images of North American Jews*, ed. Allon Gal (Jerusalem and Detroit: The Magnes Press and Wayne State University Press, 1996), 193–218; Yigal Elam, *The Jewish Agency: First Years, 1919–1931* (Jerusalem: The Zionist Library, 1990) (Hebrew).

7. Zagorodsky, *Citrus Book*, A., 103.

8. Oppenheim, I. D., "The Development of Cultural Practices in the Citrus Industry," *Hadar* 11:1 (1938): 40–42 (Hebrew); Yedidyah, *Citrus Growing*, 233.

9. Ruppin, *Agricultural Settlement*, 125–33; see also various cost calculations for planting of medium and small orchards, as worked out for potential settlers, CZA, S15/2556.

10. Hanotaiah, *Netanyah*; Idem., *Orange Groves*.

11. Yakhin Cooperative Society, *Statutes*; Idem., *Four Years*.

12. Ryerson, *Reports*, 273–74.

13. Hazen, *The Citrus Industry of Palestine*, 41–42; H. Clark Powell, "The Citrus Industry in Palestine," Government of Palestine (Department of Agriculture and Forests, Agricultural Leaflets, Series IV), *Horticulture* 9 (1928).

14. "On the Question of Congestion of Planting in Citrus Orchards," *Bulletins for Yakhin Employees*, 7 (1934), 9–13; and summation of the issue in Hazen, *Citrus Industry*, 41–42.

15. Zagorodsky, *Citrus Book*, B, 211.

16. B. Gorstein, "The Citriculture Account in Palestine," *Hasadeh* 16 (1936): 580–82; Hazen, *Citrus Industry*, 14.

17. Viteles, "Expansion of the Orange Industry"; Idem., "Survey of the Citrus Industry in Palestine," *Hadar* 11:1, 2–3 (1938): 1–43 (Hebrew).

18. Yedidyah, *Citrus Growing*, 253; Zagorodsky, *Citrus Book*, B, 40–41; A. Goldwater, *Citrus Fruit in Palestine: Their Development and Prospects* (Jerusalem: Palestine Corporation, 1927), 14.

19. Zagorodsky, *Citrus Book*, B, 40–43; Ryerson, *Reports*, 264–73. Goldwater, *Citrus Fruit*, 13–14, reports that of 42,000 dunams of orchards, 3,000 were irrigated with surface water and 3,000 by water companies. The remaining 36,000 dunams were irrigated from private wells.

20. Yedidyah, *Citrus Growing*, 253–54; I. Eisenberg, "Upper Drilling and Mixed Well," *Bustenai* 6, 45 (Feb. 27, 1935): 25–26 (Hebrew); Ibid. 46 (March 6, 1935): 25–26; Idem., *Pumping Systems for Irrigation* (Tel Aviv: Farmers' Federation, 1938) (Hebrew).

21. Zagorodsky, *Citrus Book*, B, 53–62; Yedidyah, *Citrus Growing*, 252–54; Eisenberg, *Pumping Systems*, 21–49; M. Seitz, "How to Set Up a Water-Delivery System," *Hasadeh* 14 (1934): 506–10, 607–10 (Hebrew).

22. Avitsur, *Industrial Revolution*, 102–18; Basim A. Faris, *Electric Power in Syria and Palestine* (Beirut: The American Press, 1936).

23. Farmers' Federation Archives, file 13.

24. Zagorodsky, *Citrus Book*, B, 27–62; Yedidyah, *Citrus Growing*, 251–57; Eisenberg, *Pumping Systems*; Seitz, "How to Set Up a Water-Delivery System"; Gross, "The Economic Policy of the Mandatory Government in Palestine."

25. M. Seitz, "Problem of the Water-Delivery System," *Bulletins for Yakhin Employees* 10 (Aug. 1934): 1–8; Yakhin Cooperative Association for Agricultural Contracting, Ltd., Tel Aviv [?], 1929, pages unnumbered (Hebrew and Yiddish); "Local Chronicle," *Hadar* 2, 7 (July 1929); "From the Colonies," *Bustenai* 5, 7 (June 1, 1933).

26. See Stephan Davies, *The Diffusion of Process Innovations* (Cambridge: Cambridge University Press, 1979); Gershon Feder, Richard E. Just, David Zilberman, "Adopting of Agricultural Innovations in Developing Countries: A Survey," *Economic Development and Cultural Change* 33: 2 (1985): 255–98; Edwin Mansfield, "Technical Change and the Rate of Imitation," in *The Economics of Technological Change: Selected Readings*, ed. Nathan Rosenberg (Great Britain: Penguin,1971), 284–315; Rosenberg, *Technology and American Economic Growth.*

27. Zagorodsky, *Citrus Book*, A, 98, and B, 42–43; "From the Colonies," *Bustenai* 5, 40 (Jan. 24, 1934); Ibid., 6, 22 (Sept. 12, 1934); Avitsur, *Changes*, 124; Yedidyah, *Citrus Growing*, 256; Eisenberg, "Upper Drilling," note 20 above; on resistance to new knowledge and technologies see Mokyr, *The Gifts of Athena*, 218–83.

28. Ryerson, *Reports*, 269–71, 280–82.

29. Katzparovsky, "California," 9–10. In regard to California irrigation, see, for example, Zagorodsky, *Citrus Book*, A, 95; Yedidyah, *Citrus Growing*, 259–66.

30. Zagorodsky, *Citrus Book*, A, 95–97, B, 44–49; Moshe Smilansky, "Irrigation in Citrus Orchards," *Hasadeh* 2, 10 (July 1922): 356–59 (Hebrew); Idem., "Planting a Citrus Orchard," *Hasadeh* 7 (1926): 262–69, 320–23 (Hebrew); Yakhin, *Four Years of Activities*.

31. Zagorodsky, *Citrus Book*, B, 49.

32. Hazen, *Citrus Industry*, 41.

33. Paul A. David, "The Mechanization of Reaping in the Ante-Bellum Midwest," in *The Economics of Technological Change: Selected Readings*, ed. Nathan Rosenberg (Great Britain: Penguin, 1971), 229–73; Idem., *Technical Choice, Innovation, and Economic Growth: Essays on American and British Experience in the Nineteenth Century* (Cambridge: Cambridge University Press,1975); Davies, *Diffusion of Innovations*.

34. Yedidyah, *Citrus Growing*, 259.

35. Zagorodsky, *Citrus Book*, A, 88–89.

36. For a summary of the issue, see ibid., 88; Hazen, *Citrus Industry*, 41.

37. Zagorodsky, *Citrus Book*, A, 67–70; Hazen, *Citrus Industry*, 41; Yedidyah, *Citrus Growing*, 97, 156–59.

38. Moshe Guthilf, "Development of the Treatment of Citrus Fruit in the Past and Its Prospects for the Future," *Hadar* 11, 1 (Feb. 1938): 53–54 (Hebrew).

39. Hazen, *Citrus Industry*, 13–14; Yedidyah, *Citrus Growing*, 417–24; Guthilf, "Development of the Treatment of Citrus Fruit"; Israel State Archives, Bloc 7, Container 675, File H/64/21/8, "Interim Report of the Sub-Committee on Transport of the Citrus Fruit Transportation Committee," Jan. 12, 1936.

40. The discussion is based on Bavli, Y., "Our Packing Methods in the Past, the Present, and the Future," *Hadar* 7, 4–5 (Feb. 1934): 99; Guthilf (note 38 above); Yedidyah, *Citrus Growing*, 416–36; Zagorodsky, *Citrus Book*, B, 63–73; M. Goldberg, "Simple Efficiencies that Yield Important Results," *Bustenai*, 10, 27 (Nov. 2, 1938); Ibid., 6, 5 (May 16, 1934); Ryerson, *Reports*, 288–94, Viteles, *Expansion*, 43; Idem., *Survey*, 23–25; Hazen, *Citrus Industry*, 13–16. On the packing house affair and the futile attempt to reactivate the facility, see CZA, S1/680, and Fruit Commission, *Report*.

41. The discussion is based on Ryerson, *Reports*, 302–303; Yosef Braverman, "What Shall We Do with the Reject Fruit?" *Hadar* 3, 1 (Feb. 1930): 6–8; Idem., "The Problem of Unshippable Fruit *(Berara)* and Byproducts in Palestine," *Hadar* 8, 2 (March 1935): 85–87; Idem., "The Citrus Byproducts Industry," *Hadar* 11, 1 (Feb. 1938): 57–58; *Citrucine, Basic Citrus Fruit for Industry in Palestine*, Tel Aviv, 1937 (Hebrew); Yedidyah, *Citrus Growing*, 408–15; Rokach, *Tales of the Jaffa Orange Groves*, 153–54; Hazen, *Citrus Industry*, 21–23.

42. Gurevich and Gerz, *Jewish Agricultural Settlement*, 101; Hazen, *Citrus Industry*, 34, 71, 83.

43. Gurevich and Gerz, *Jewish Agricultural Settlement*, 84, 101; Yedidyah, *Citrus Growing*, 36–42, 115–55; Hazen, *Citrus Industry*, 22, 37, 83–84; Ryerson, *Reports*, 264; Moshe Smilansky, "Our Agriculture on the Eve of 1929/30," *Bustenai* 1, 29 (Oct. 30, 1929): 2.

44. A. Friedman, "Varieties of Citrus Other than Shamouti and Seedless Marsh Grapefruit That Can Be Grown Commercially in Palestine," *Hadar* 6, 3–4 (April 1932): 77–79; Zagorodsky, *Citrus Book*, B, 74–76; Yedidyah, *Citrus Growing*, 37–42, 124–25; Isaac Rokach, "The Palestine Grapefruit and Its Dissemination," *Hadar* 3, 3 (March 1930): 58–59; Ryerson, *Reports*, 307; Hazen, *Citrus Industry*, 71–88.

45. The discussion is based on Fellman, *Pardess Society*, 73–75; Rokach, *Tales of the Jaffa Orange Groves*, 248–49; Y. Oppenheim, "Problems of Shipping Fruit to Foreign Markets," *Hadar* 9, 4 (April 1936): 87–88; Yedidyah, *Citrus Growing*, 355–407; Yitzhak Elazari-Volcani, "Scientific Research in Citrus Groves," *Hadar* 5, 3–4 (March-April 1932): 81–83; Shimon Bodenheimer, "A Decade of Entomological Research in Palestine Citrus Orchards," *Hadar* 11, 1 (Feb.1938): 43–47; Yisrael Reichert, "A Decade of Investigation of Citrus Diseases in Palestine," ibid., 59–61; M. T. Dow, "Citriculture and the Government," ibid., 23–26; Zagorodsky, *Citrus Book*, B, 104–202; Hazen, *Citrus Industry*, 45–47; Ryerson, *Reports*, 285–88; CZA, S15/2556. A detailed plan for citrus research with assistance from the Government and the Jewish Agency appears in Israel State Archives, Bloc 2 (Chief Secretariat), A/39/34.

46. Viteles, *Expansion*, 50–51; Viteles, *Survey*, 8.

47. Based on statistics in Gurevich and Gerz, *Jewish Agricultural Settlement*, 84.

48. Overland and maritime shipping arrangements at various periods of time are discussed in Ryerson, *Reports*, 294–96; Hazen, *Citrus Industry*, 17–20; Isaac Rokach, "Jaffa Oranges in the 1932/33 Season," *Hadar* 6, 4 (April 1933): 106–108; Idem., "Citrus Industry and Shipping," *Palestine and Middle East Economic Magazine* 10, 6 (June 1938): 234–37, 245; Idem., "Citrus Shipping in 1938/39," ibid. (May 1939): 115–18, 136.

CHAPTER SIX. TECHNICAL INNOVATIONS IN ARAB CITRICULTURE

1. Ze'ev Abramowitz and Isaac Guelfat, *The Arab Economy in Palestine and in the Middle-Eastern Countries* (Tel Aviv: Hakibbutz Hameuchad, 1944), 1–130 (Hebrew); Abraham Cohen, *The Economy of the Arab Sector in Palestine during the British Mandate Period* (Givat Haviva: The Institute of Arab Studies, 1978), 7–26 (Hebrew); Joseph Vashitz, *The Arabs in Eretz Israel* (Merhavia: Sifriat Ha-Poalim, 1947), 13–134 (Hebrew); Yaacov Shimoni, *The Arabs of Palestine* (Tel Aviv: Am Oved, 1947), 157–205 (Hebrew); Kamen, *Little Common Ground*; Issa Mustafa Alami, *Some Aspects of the Development of the Palestinian Peasant Economy and Society, 1920–1939* (PhD dissertation, Edinburgh, Edinburgh University, 1984), 468–96; Metzer and Kaplan, *Jewish and Arab Economies*, 19–53, 137–72; Metzer, *Divided Economy*, 138–54. The studies of Metzer and Kaplan and of Metzer *(Divided Economy)*, in my opinion, lend Arab citriculture the right degree of importance in the Arab economy and agriculture.

2. Gurevich and Gerz, *Jewish Agricultural Settlement*, 84*.

3. Ibid., 84* and 35*, Table 23; Metzer and Kaplan, *Jewish and Arab Economies*, Table I-1, and discussion on 167–72.

4. Cohen, *Arab Sector*, 7–26; Kamen, *Little Common Ground*, 183, 213–14.

5. All the following quotations and summaries are culled from reports by Lubmann Haviv, 1932, in CZA, S25/7621.

6. For information of Lubmann Haviv, see David Tidhar, *Encyclopaedia of the Pioneers and Builders of the Yishuv, Vol. 17* (Tel Aviv: 1968), 5218 (Hebrew).

7. See Abramowitz and Guelfat, *Arab Economy in Palestine*, 1–130; Kamen, *Little Common Ground*, 135–73; Kenneth W. Stein, *The Land Question in Palestine 1917–1939* (Chapel Hill and London: University of North Carolina Press, 1984).

8. In regard to them, see Amit, *Jewish Settlement Activity*, 231–47, 287–312.

9. For information about Gardinger, see Rokach, *Tales of the Jaffa Orange Groves*, 232–34; Amit, *Jewish Settlement Activity*, 232–41, 248–312.

10. See note 5 above.

11. Amit, *Jewish Settlement Activity*, 146–70.

12. Smilansky, "Jewish Colonization and the Fellah."

13. Shimoni, *Arabs of Palestine*, 233–34.

14. "From the Colonies," *Bustenai* 9: 8 (1937/8): 28–29.

15. "In the Arab Economy of Palestine," *Mishar Ve-Ta'asiyah* 10: 14–15 (Nov. 1932): 21 (Hebrew); 17 (Dec. 1932): 10; in regard to these families, see Shimoni, *Arabs of Palestine*, 103–104, 228–34.

16. Ben Yisrael (Ze've Smilansky), "In Our Neighbors' Villages," *Bustenai* 6: 15–16 (1934): 16–17, 25–27 (Hebrew). All the following quotations and summaries are culled from this article.

CHAPTER SEVEN. PURSUIT OF PROFIT

1. CZA, A323/435, *Pardess Society Book of Minutes*, Part 2, 1910 (pages unnumbered) (Hebrew).

2. Moshe Smilansky, "Our Agriculture on the Eve of 1930/31," *Bustenai* 2: 25 (Sept. 26, 1930): 1–3. One may get an idea about prices during the reasonable-profitability period from Isaac Rokach's annual reports in *Hadar* about the results of the various seasons. Additional reports were provided by Smilansky in *Bustenai* and by Tolkowsky, Viteles, and others in various forums. Reports by Pardess, as we have seen, are excessively general and do not provide an appropriate level of price details.

3. Y. (Yosef) Sapir, "The Citrus Industry from the Investor's Perspective," *Hadar* 5: 3–4 (March-April 1932): 69–71.

4. A. Polani, "The 1936/37 Palestine Citrus Season," *Hadar* 10: 5 (May 1937): 57–59 (Hebrew); Moshe Smilansky, "The Fate of Citriculture," *Bustenai* 8: 47 (March 17, 1937): 3–4 (Hebrew); Idem., "Our Citriculture," *Bustenai* 9: 20 (Aug. 25, 1937): 1–2 (Hebrew); "Rentability of Citriculture in the 1936/37 Season," *Bulletin of the Institute for Economic Research* 2: 1–2 (Jan.-Feb. 1938): 13–15 (Hebrew); S. Tolkowsky, "The 1936–37 Citrus Season and the Future Outlook," *Palnews: Economic Annual of Palestine* 3 (1937): 95–100.

5. "Rentability of Jewish Citriculture in the 1937/38 Season," *Bulletin of the Institute for Economic Research* 3: 1–2 (Jan.-Feb. 1939): 14–16 (Hebrew); Moshe Smilansky, "Summing Up the Season," *Bustenai* 10: 3 (May 11, 1938): 1–2 (Hebrew); Isaac Rokach, "The 1937/38 Palestine Citrus Season," *Hadar* 11: 5 (May 1938): 111–12 (Hebrew).

6. Gross, "Economic Policy of the Mandatory Government"; Hazen, *Citrus Industry*, 56; Max Beloff, *Wars and Welfare: Britain 1914–1945* (London: Arnold, 1984), 188–95; Ian-William Gaskin, *Economic Aspects of the Citrus Industry in Palestine 1918–1948* (Liverpool: Liverpool University, 1985), 9–17.

7. Shulamit Mashke, *The Problem of Marketing the Palestinian Oranges in Europe* (Jerusalem: 1937), especially 14–17 (the quotation is from 17).

8. Viteles, *Expansion*, 10–67; Yedidyah, *Citrus Growing*, 3–29; Mashke, *Marketing*; Hazen, *Citrus Industry*, 1, 56–67.

9. This discussion is based mainly on reports appeared in the Jewish press of the time. For detailed references see Nahum Karlinsky, *Citrus Blossoms: Jewish Entrepreneurship in Palestine 1890–1939* (Jerusalem: Magnes Press, 2000), 227 (Hebrew). See

also: Yedidyah, *Citrus Growing*, 11–13, 377–78; Chaim Mendelsohn, *International Trade in Oranges: Competition for Export Markets* (Unpublished PhD dissertation, University of California, Berkeley, 1958), 104–10, 264–88; Raymond Carr, *Spain: 1808–1939* (Oxford: Clarendon, 1966), 603–94.

10. Ryerson, *Report*, 298; Hazen, *Citrus Industry*, 54–55.

11. For detailed references see Karlinsky, *Citrus Blossoms*, 229; Shmuel Tolkowsky archives, CZA, A248/5.

12. Viteles, *Expansion*, 60–61; Idem., *Survey*, 30–37; Hazen, *Citrus Industry*, 57–61; Mendelsohn, *Oranges*, 70–133, 288–98; Poshter, *Cooperative Societies*, 19–41, Tables 7 and 9; "Shipments of Pardess Syndicate and Ha-Haklai in the 1938/39 Season," CZA, A323/101; "Pardess Syndicate Goes Into Action," *Hadar* 12: 4 (May 1939): 76.

13. Mashke, *Marketing*; "Shipments of Pardess Syndicate and Ha-Haklai . . . ," CZA, A323/101; Hazen, *Citrus Industry*, 53–77; CZA, A323/75.

CHAPTER EIGHT. THE BRITISH MANDATE GOVERNMENT'S POLICY TOWARD THE CITRUS INDUSTRY

1. Gross, "Economic Policy of the Mandatory Government."

2. See references in chapter 7, Note 6; CZA, A323/82.

3. M. T. Dow (director of the Department of Agriculture), "Citriculture and the Government," *Hadar* 11: 1 (Feb. 1938), and references in previous note.

4. E. R. Sawer (Director of Agriculture and Forests, Palestine Government), "Introduction," *Hadar* 3: 7–8 (July–August 1930): 155–56; Idem., *A Review of the Agricultural Situation in Palestine* (Palestine: Government of Palestine, Department of Agriculture and Fisheries, 1922); Department of Agriculture, Annual Reports; minutes of citrus committees in CZA, S15/1817A3; various reports in the archives of Isaac Rokach, CZA, A323; and in the personal archives of Zalman Jacobsohn, who was posted to the committee ex officio; Dow, "Citriculture"; and I. Rokach, "The Orange Industry and the Government," *The Palestine Citrograph* 1: 1 (Feb. 1928): 5–6; editorials, ibid. 1, 10 (Nov. 1928): 3–4.

5. This section is based on the following: *Report of the Fruit Commission*, 22–23; Shmuel Tolkowsky, "About the Ordinance concerning Shipment of Fruit from Palestine, *Mishar Ve-Ta'asiyah* 5: 18 (Nov. 1927): 290; R. O. Williams, "A Decade of Citrus Inspection in Palestine," *Hadar* 11: 1 (Jan. 1938): 50–52 (Hebrew); Dow, "Citriculture"; Isaac Rokach, "Remarks on the Inspection Regulations for the Citrus Grower, 1936," *Hadar* 9: 9 (Aug. 1936): 159–60 (Hebrew); Y. B. (Yosef Braverman), "On the Agenda," *Hadar* 10: 6 (June 1937): 77; I. R. (Isaac Rokach), "On the Agenda," *Hadar* 10: 9 (Sept. 1938): 121 (Hebrew).

6. Dow, "Citriculture"; Hazen, *Citrus Industry*, 28–29; E. J. White, "The First Advertising Campaign for Jaffa Citrus: 'Jaffa for Juice'—A Slogan That Will Live for Years," *Hadar* 7: 4–5 (April–May 1934): 25; Gordon Boggon, "Notes on Jaffa Citrus Advertising 1934," *Hadar* 7: 3 (March 1934): 65–66; Idem., "The Great Jaffa Citrus Publicity Scheme," *Hadar* 8: 4 (April 1935): 110–12.

CHAPTER NINE. ATTEMPTS TO ESTABLISH A CARTEL

1. United States Department of Agriculture, Bulletin no. 1234, *Organization and Development of a Cooperative Citrus-Fruit Marketing Agency*, Washington, DC, 1924; William W. Cumberland, *Cooperative Marketing: Its Advantages as Exemplified in the California Fruit Growers Exchange* (Princeton: Princeton University Press, 1917); *California History* LXXIV: 1 (Spring 1995): A special volume on *Citriculture and Southern California*; Karlinsky, "California Model"; Steven Stoll, *The Fruits of Natural Advantage: Making the Industrial Countryside in California* (Berkeley, Los Angeles, and London: University of California Press, 1998); CZA, A323/435; Fellman, *Pardess Society*, 65. For attempts by banks to unify the industry, see Nachum T. Gross, "Die Deutsche Palaestina-Bank 1897–1914," *Zeitschrift fuer Unternehmensgeschichte* 33:3 (1988): 149–77.

2. CZA, A249/26; CZA, A324/435; Gross, "Die Deutsche Palaestina-Bank."

3. E .R. Sawer, "Introduction," in Viteles, "Proposal to Organize the Orange Industry in Palestine," *Hadar* 3: 7–8 (June–July 1930): 3–5; Viteles's initiative was accompanied by nonpublic meetings with the leaders of the Jewish citrus industry to organize a "Palestine Fruit Growers Exchange" along the Californian model. See: The New York Public Library, Manuscripts and Archives Division, PEC collection, Box 91. On the progressive worldview of the Brandeis-Mack group, to which ideas Viteles adhered, and on its economic initiatives in Palestine, see: Allon Gal, "Brandies' View on the Upbuildng of Palestine," *Studies in Zionism* 6 (1982): 211–40.

4. Ryerson, *Reports*, 297.

5. Mancur Olson, *The Logic of Collective Action: Public Goods and the Theory of Groups* (New York: Schocken, 1965).

6. Ryerson, *Reports*, 296–98; Viteles, "Expansion," 69–73; Report of Fruit Commission; manuscript of lecture by Rokach to Petah Tikva growers in 1934, titled "Cooperation in Citriculture," and quotation, ibid., see CZA A323/409.

7. Personal archives of Zalman Jacobsohn.

8. Fellman, *Pardess Society*, 32. On the diffusion of cooperative institutions around the world at that time see: C. R. Fay, *Co-operation at Home and Abroad: A Description and Analysis* (London: King and Son, 1936–1939).

9. CZA A323/435.

10. Fellman, *Pardess Society*, 32–50; Mazie archives; CZA, A/323/435, *Pardess Society Book of Minutes*; CZA, A249/7, A249/37C; Pardess Society, *Statutes* (before 1914); Viteles, "Expansion," 68.

11. Nawratzki, *Kolonisation*, 174; CZA, A323/435, *Pardess Society Book of Minutes*; Fellman, *Pardess Society*, 32–67, 72; Tolkowsky, "Organizing the Citrus Trade in Palestine," *Hadar* 13: 3 (March 1940), 45.

12. Viteles, "Expansion," 67–69; CZA, A323/44, Report of the General Manager of the Pardess Cooperative Society, I. Rokach, Tel Aviv, 1930.

13. CZA, A323/435; Pardess Society, *Statutes* (1910–1930); Viteles, "Expansion," 67–73; Pardess, reports for 1920/21–1925/26 seasons.

14. Pardess, reports for 1926/27–1931/32 seasons, report for 1938/39 season, and summarizing table, CZA, A/323/44, 12.

15. Viteles, "Survey," 11. Viteles also noted in this article that one Arab cooperative society was recorded with the Registrar of Cooperative Societies at the time, but it was not active.

16. Registrar of Cooperative Societies in Palestine, *Report by the Registrar of Cooperative Societies in Palestine on the Development During the Years 1921–1937* (Jerusalem: Government Printing Press, 1938); Isaac Rokach, "Achievements of the Citrus Industry in the Past Two Years," *Bustenai* 6: 42 (Feb. 6, 1935): 30–37.

17. Shapira, *Futile Struggle*, 102–36, 345–46; Avraham Sela, "The Disturbances at the Wailing Wall (1929): A Turning Point in Jewish-Arab Relations?" in *Jerusalem in Zionist Vision and Realization*, ed. Hagit Lavsky (Jerusalem: Zalman Shazar Center for Jewish History, The Center for the Study of Zionism and the Yishuv, Hebrew University of Jerusalem, 1989), 261–78 (Hebrew); Israel Kollat et al., "Discussion: The 1929 Disturbances as a Turning Point," in *Jerusalem in Zionist Vision and Realization*, ed. Hagit Lavsky (Zalman Shazar Center for Jewish History, The Center for the Study of Zionism and the Yishuv, Hebrew University of Jersalem, 1989), 407–30 (Hebrew).

18. Rokach, *Tales of the Jaffa Orange Groves*, 13–54; A. Arnsdorf, "The Development of Citrus Shipments from Palestine during the Past Decade," *Hadar* 11: 3–4 (March–April 1938): 89–93.

19. Shmuel Tolkowsky, "Organizing Orange Trading in Palestine," *Hadar* 13: 3 (Feb. 1940): 46; Shmuel Tolkowsky archives in CZA, A248, files 5, 11, 12; Gershon Gera, *The Unknown Generous Benefactor* (Tel Aviv: Modan, 1984) (Hebrew); interview with Ada Auerbach, daughter of Shmuel Tolkowsky, June 11, 1992; Isaac Rokach archives, CZA, A323/354.

20. Based on Israel State Archives, Bloc 7, Container *mem*/675, files H/64/21/7,8; CZA, S15/2817A; S15/2817/A3; Rokach, *Tales of the Jaffa Orange Groves*; Shimoni, *Arabs of Palestine*, 206–40.

21. This discussion is mainly based on various reports appeared in the Jewish press. See for example Moshe Smilansky, "Our Citriculture," *Bustenai* 4: 1–2 (April 20, 1932): 3–5; Idem., "The Citrus Department," *Bustenai* 3: 21 (Sept. 2, 1931): 3–4.

22. Fellman, *Pardess Society*, 109; CZA, L51/374; CZA, A323/44; Rokach, *Tales of the Jaffa Orange Groves*, 123–24, 150–54, 241–43; "Our Mission," *Hadar* 1: 1 (Feb. 1928): 3–4; Pardess reports.

23. CZA, A323/77; CZA, A323/483, "Plan for Centralization of Packing at the Pardess Society," April 29, 1932.

24. CZA, A323/44, A323/201.

25. Rokach diaries: CZA, 201, 205, 255, 354, A323/77.

26. CZA, A323/354.

27. CZA, A323/77.

28. Moshe Smilansky, "United Cooperative," *Bustenai* 6: 43 (Feb. 6, 1935): 5–7; Rokach, CZA, A323/354.

29. CZA, A323/354, quoted in Viteles, "Survey," 7–8.

30. The discussion that follows is based on references appeared in Karlinsky, *Citrus Blossoms*, 274 and: "General Actions Taken ahead of Citrus Marketing in the 1939/40 Season by or with the Active Participation of the Union," *Bustenai* 11: 19 (Aug. 23, 1939): 13; Fellman, *Pardess Society*, 133–75; A. Poshter, *The Development of the Citrus Marketing Cooperative Societies in the 1937/38 Season* (The Anglo-Palestine Bank, March 1939);Eliezer Ziegfried Hoofien, Zvi Viteles, and Mark Yaffe, *Recommendations for the Unification of the Jewish Citrus Industry's Cooperatives* (Tel Aviv: 1938) (Hebrew); Final Report (Internal) of the 'Committee of Three,'" March 14, 1939, in Anglo-Palestine Bank archives, CZA, and additional material on the union, CZA, L51/1909; and CZA, L51/1911, L61/1912; Rokach archives, CZA, A323; interview with 'Uzi Komarov, Feb. 23, 1993.

31. CZA, A323/354.

32. Hoofien, Viteles, and Yaffe, *Recommendations*; "Proposal for Unification of the Citrus Growers' Organizations," *Hadar* 11: 8 (July 1938): 171–73, 178; E. Hoofien, "Unifying the Cooperatives," *Bustenai* 10: 43 (Feb. 15, 1939); CZA, L51/1909, L51/1911, L51/1912.

33. "General Actions Taken ahead of Citrus Marketing" (above, note 30); Isaac Rokach, "Problems and Prospects of Citriculture," *Hadar* 13: 6 (June 1940): 105–108 (quotation from 105); Idem., "The Proposed Citrus Control Board," *Hadar* 13 (April 1940); Gross, "Economic Policy of the Mandatory Government"; Rokach, *Tales of the Jaffa Orange Groves*, 136–43; Fellman, *Pardess Society*, 157–93.

Bibliography

ARCHIVES

Israel State Archives
Central Zionist Archives (CZA)
Labor Archives—The Lavon Institute for Labor Research
Farmers' Federation Archives
Rehovot Municipal Archives
Yad La-banim Archives—Petah Tikva
Aharon Meir Mazie Archives
Zalmann Jacobssohn Archives
Shmuel Holtzmann Archives
New York Public Library, Palestine Economic Corporation Archives

INTERVIEWS

Ada Auerbach
Deborah Dimant
Daniel Jacobsohn
Hillel Jacobsohn
Imanuel Jacobsohn
'Uzi Komarov
Izhar Smilansky
Rinah Smilansky
Dan Tolkowsky
Ahuvah Zmora
Leah Zuker

PERIODICALS

Bustenai
Hadar
Ha-Haklai
Hasadeh
Kuntres
Mis'har Ve-Ta'asiyah
Pal News: Economic Annual of Palestine
Palestine and Middle East Economic Magazine

PUBLISHED REFERENCES—PRIMARY AND SECONDARY

Aaronsohn, Aaron. *Citrus Fruit around the Globe as It Seems from the Palestinian Grower's Perspective*. Jaffa: The Zionist Organization, Agricultural Experimental Station, 1915 (Hebrew).

Aaronsohn, Aaron, and Selig Soskin. "Die Orangengaerten von Jaffa." *Der Tropenpflanzer* 4 (1902): 341–61.

Aaronsohn, Ran. "Stages in the Development of the Settlements of the First Aliyah." In *The First Aliyah*, ed. Mordechai Eliav. Jerusalem: Yad Izhak Ben-Zvi and Israel Ministry of Defence Press, 1981, 25–84 (Hebrew).

———. "Baron Rothschild and the Initial Stage of Jewish Settlement in Palestine (1882–1890): A Different Type of Colonization?," *Journal of Historical Geography* 19:2 (1993): 142–56.

———. "Vines and Wineries in The Jewish Colonies: Introducing Modern Viticulture into Nineteenth-Century Palestine," *Studies in Zionism* 14:1 (1993): 32–51.

———. "The Beginnings of Modern Jewish Agriculture in Palestine: 'Indigenous' Versus 'Imported,'" *Agricultural History* 69:3 (1995): 438–53.

———. "Baron Rothschild's Officials, 1882–1890." *Cathedra* 74 (1994): 157–78 (Hebrew).

———. "Settlement in *Eretz Israel*—A Colonialist Enterprise? 'Critical' Scholarship and Historical Geography." *Israel Studies* 1:2 (1996): 214–29.

———. *Rothschild and Early Jewish Colonization in Palestine*. Jerusalem: Magnes Press, 2000.

Abramowitz, Ze'ev, and Isaac Guelfat. *The Arab Economy in Palestine and in the Middle-Eastern Countries*. Tel Aviv: Hakibbutz Hameuchad, 1944 (Hebrew).

Agriculture, Annual Reports. *Government of Palestine—Department of Agriculture: Annual Reports*. Jerusalem: Government of Palestine—Department of Agriculture, various years.

Ahad Ha-'Am (Asher Ginsberg). *Collected Writings of Ahad Ha-'Am (Kol kitvei Ahad Ha-'Am)*. Tel Aviv: Dvir, 1957 (Hebrew).

Aharoni, Yair. "Some Notes on Entrepreneurship and Entrepreneurs." *Economic Quarterly* 14 (1967): 286–74 (Hebrew).

Alami, Issa Mustafa. *Some Aspects of the Development of the Palestinian Peasant Economy and Society, 1920–1939*. PhD dissertation, Edinburgh, Edinburgh University, 1984.

Almog, Oz. *The Sabra: The Creation of the New Jew*. Berkeley: University of California Press, 2000.

Amikam, Bezalel. *First Tillings: The Jewish Agricultural Settlements in the 1930s*. Jerusalem: Yad Izhak Ben-Zvi, 1980 (Hebrew).

Amit, Irit. "American Jewry and the Settlement of Palestine: Zion Commonwealth Inc." In *The Land that Became Israel: Studies in Historical Geography*, ed. Ruth Kark. New Haven, London, and Jerusalem: Yale University Press and The Magnes Press, 1989, 251–71.

———. "The Shaykh Muwannis Land Dispute and Its Results." In *Studies in the Historical-Settlement Geography of Palestine*, B, ed. Yossi Katz, Yehoshua Ben-Aryeh, and Yehoshua Kaniel. Jerusalem: Yad Izhak Ben-Zvi, 1991, 167–84 (Hebrew).

———. "The Place of Great Capital in Shaping the Sharon in the Second Half of the Second Decade of the Twentieth Century." *Studies in the Geography of Israel* 14 (1993): 122–41 (Hebrew).

———. *Jewish Settlement Activity in the Southern Sharon 1918–1929: National Capital and Private Enterprise in the Shaping of the Southern Sharon*. PhD dissertation, the Hebrew University of Jerusalem, Jerusalem, 1993 (Hebrew).

———. "Private Jewish Investors in Palestine in the 1920s." In *The Mosaic of Israeli Geography*, ed. Yehuda Grados and Gabriel Lipshitz. Beersheva: Ben-Gurion University of the Negev Press, 1996, 451–59.

Aner, Zeev. "Holzmann: The Man Who Gave His Name to the Etzion Bloc." In *The Etzion Bloc from Its Inception to 1948 (Idan 7)*, ed. Mordechai Naor. Jerusalem, 1986, 53–62 (Hebrew).

Apfelbaum, Meir. "Citrus Growing in Palestine." *HaHaklai* 1 (1912/13): in segments, 1–290 (Hebrew).

Avitsur, Shmuel. "The First Project for the Intensive Exploitation of the Yarkon Waters (The Frangija-Navon Scheme of 1893)." *Museum Haaretz Bulletin* 6 (1964): 80–88.

———. *The Rise and Fall of the Port of Jaffa*. Tel Aviv: Milo, 1972 (Hebrew).

———. *Changes in Agriculture of Eretz Israel 1875–1975*. Tel Aviv: Milo, 1978 (Hebrew).

———. "The Contribution of Early Petah Tiqva to the Agricultural and Industrial Development of Eretz Israel (1878–1918)." *Cathedra* 10 (1979): 129–41 (Hebrew).

———. "A Pioneer of Modern Industry in Eretz Israel: The History of the L. Stein Factory." *Cathedra* 15 (1980): 69–94 (Hebrew).

———. *The Banks of the Yarkon River (1947)*. Tel Aviv: Reshafim, 1980 (Hebrew).

———. "Agriculture, Handicrafts, and Light Industry during the Period of the First Aliyah." In *The First Aliyah*, ed. Mordechai Eliav. Jerusalem: Yad Izhak Ben-Zvi and Israel Ministry of Defence Press, 1981, 225–44 (Hebrew).

———. *Inventors and Adaptors: The Persons Who Revolutionized the Production Process in Eretz Israel*. Edited by Rehavam Zeevi and Genia Doron. Jerusalem and Tel Aviv: Yad Izhak Ben-Zvi and Haarezt Museum, 1986 (Hebrew).

———. *The Industrial Revolution in Eretz Israel in its Energetic Processes*. Tel Aviv: Hakibbutz Hameuwhad, 1989 (Hebrew).

Barkai, Haim. *Adam Smith in the Context of Our Times*. Jerusalem and Tel Aviv: The Bialik Institute and the Open University, Israel, 1995 (Hebrew).

Bassok, Ido. *In the Shadow of Orchards on a Leprous Soil: Studies in the Literary and Essayitic work of Moshe Smilansky and Uri Zvi Grinberg*. Tel Aviv: Hakibbutz Hameuchad, 1996 (Hebrew).

Beenstock, Michael, Jacob Metzer, and Sanny Ziv. "Immigration and Jewish Economy in Mandatory Palestine." *Research in Economic History* 15 (1995): 149–213.

Beilin, Yossi. *Roots of Israeli Industry*. Jerusalem: Keter, 1987 (Hebrew).

Bein, Alex. *History of Zionist Settlement: From Herzl to the Present*. Ramat-Gan: Masada, 1970 (Hebrew).

Beloff, Max. *Wars and Welfare: Britain 1914–1945*. London: Arnold, 1984.

Ben-Artzi, Yossi. "Development of the First Aliya Colonies and the Establishment of New Colonies during the Second Aliya Period." In *The Second Aliya: Studies*, ed. Israel Bartal. Jerusalem: Yad Izhak Ben-Zvi, 1997, 104–34 (Hebrew).

———. *Early Jewish Settlement Patterns in Palestine, 1882–1914*. Jerusalem: The Magnes Press, 1997.

Ben-Artzi, Yossi, and Abrahahm Labes. "Netanyah's First Plan." *Ofakim Be-Geographiyah* 6 (1982): 65–68 (Hebrew).

Ben-Aryeh, Yehoshua. "Geographic Aspects of the Development of the First Jewish Settlements in Palestine." In *The First Aliyah*, ed. Mordechai Eliav. Jerusalem: Yad Izhak Ben-Zvi and Israel Ministry of Defence Press, 1981, 85–96 (Hebrew).

Ben-Avram, Baruch. "The Emergence of the Kvutza from the Aspiration of the Pioneers of the Second Aliya for Self-Employment." *Cathedra* 18 (1981): 118–23 (Hebrew).

Ben-Avram, Baruch, and Henry Near. *Studies in the Third Aliyah: Image and Reality*. Jerusalem: Yad Izhak Ben-Zvi, 1995 (Hebrew).

Ben Gurion, David. *From Class to Nation: Reflections on the Goals and Policies of the Workers' Movement*. Tel Aviv: Davar, 1933 (Hebrew).

Ben-Porat, Amir. *The Bourgoise: The History of the Israeli Bourgoises*. Jerusalem: Magnes Press, 1999 (Hebrew).

Berlovitz, Yaffah. "Moshe Smilansky—'Benei Arav: The Patronizing Stories." *Iyunim Bitkumat Israel* 4 (1994): 400–21 (Hebrew).

———. *Inventing a Land, Inventing a People*. Tel Aviv: Hakibbutz Hameuchad, 1996 (Hebrew).

Biger, Gideon. *Crown Colony or National Homeland?: British Influence upon Palestine, 1917–1930—A Geo-Historical Analysis*. Jerusalem: Yad Izhak Ben-Zvi, 1983 (Hebrew).

Bonacich, Edna. "A Theory of Ethnic Antagonism: The Split Labor Market." *American Sociological Review* 37 (1972): 547–59.

———. "The Other Side of Ethnic Entrepreneurship: A Dialogue with Waldinger, Aldrich, Ward, and Associates." *International Migration Review* 7 (1993): 685–92.

Bonacich, Edna, and John Modell. *The Economic Basis of Ethnic Solidarity: Small Business in the Japanese American Community*. Berkeley: University of California Press, 1980.

California History. LXXXIV:1 (Spring 1995): A special volume on *Citriculture and Southern California*.

Carr, Raymond. *Spain: 1808–1939*. Oxford: Clarendon, 1966.

Clark Powell, H. "The Citrus Industry in Palestine." Government of Palestine. Department of Agriculture and Forests, Agricultural Leaflets, Series IV, *Horticulture* 9 (1928).

Clemence, Richard V., and Francis S. Doody. *The Schumpeterian System*. Cambridge, MA: Addison-Wesley, 1950.

Cohen, Abraham. *The Economy of the Arab Sector in Palestine during the British Mandate Period.* Givat Haviva: The Institute of Arab Studies, 1978 (Hebrew).

Cumberland, William W. *Cooperative Marketing: Its Advantages as Exemplified in the California Fruit Growers Exchange*. Princeton: Princeton University Press, 1917.

David, Paul A. "The Mechanization of Reaping in the Ante-Bellum Midwest." In *The Economics of Technological Change: Selected Readings*, ed. Nathan Rosenberg. Great Britain: Penguin, 1971, 229–73.

———. *Technical Choice, Innovation, and Economic Growth: Essays on American and British Experience in the Nineteenth Century*. Cambridge: Cambridge University Press, 1975.

Davies, Stephan. *The Diffusion of Process Innovations*. Cambridge: Cambridge University Press, 1979.

Department of Agricultural Settlement—Zionist Organization: Reports for the 14th, 15th, and 16th Zionist Congresses. Jerusalem and London: Department of Agricultural Settlement, 1925–1929 (Hebrew).

Diskin, Mordechai. *The Words of Mordechai.* Jeruslaem: 1889 (Hebrew).

Dom, Rissa. "The Character of the Arab in Moshe Smilansky's Literary Works." *Iton 77* 84–85 (1987): 95–97 (Hebrew).

Drori, Yigal. "The Second Aliya and the Struggle for Jewish Labour as Reflected in the General Eretz Israel Press." *Cathedra* 2 (1977): 69–80 (Hebrew).

———. *The "Civil Circles" in the Yishuv during the 1920s*. Tel Aviv: Mif'alim Universitaim, 1990 (Hebrew).

Eckstein, Alexander. "Introduction." In *Comparison of Economic Systems; Theoretical and Methodological Approaches*, ed. Alexander Eckstein. Berkeley: University of California Press, 1971, 1–23.

Eisenberg, I. *Pumping Systems for Irrigation.* Tel Aviv: Farmers' Federation, 1938 (Hebrew).

Elam, Yigal. *The Jewish Agency: First Years, 1919–1931*. Jerusalem: The Zionist Library, 1990 (Hebrew).

Faris, Basim A. *Electric Power in Syria and Palestine*. Beirut: The American Press, 1936.

Farmers Federation—Citricultre Department. Tel Aviv: Jewish Farmers Federation—Citricultre Department, various years (Hebrew).

Farmers Federation. *Annual Reports of the Jewish Farmers Federation in Palestine*. Tel Aviv: Jewish Farmers Federation, 1931–1940 (Hebrew).

Fay, C. R. *Co-Operation at Home and Abroad: A Description and Analysis*, two volumes. London: King and Son, 1936–1939.

Feder, Gershon, and Gerald T. O'Mara. "Farm Size and the Diffusion of Green Revolution Technology." *Economic Development and Cultural Change* 30:1 (1981): 59–76.

Feder, Gershon, Richard E. Just, and David Zilberman. "Adopting of Agricultural Innovations in Developing Countries: A Survey." *Economic Development and Cultural Change* 33:2 (1985): 255–98.

Fellman, Aharon. *Fifty Years of the Pardess Society, Fifty Years of Jewish Citriculture: 1900–1950*. Tel Aviv: Pardess Society, 1950 (Hebrew).

Fellman, Aharon Leib. *Ma'ayan Ganim: Rules for Planting Gardens and Citrus Orchards in the Holy Land*. Jerusalem: 1891 (Hebrew).

Fellman, Avraham. *The Pioneers of Jewish Citriculture in Palestine (the Fellman Family)*. Tel Aviv: Fellman, 1940 (Hebrew).

Fieldhouse, David Kenneth. *The Colonial Empires: A Comparative Survey from the Eighteenth Century*. London: Weidenfeld and Nicolson, 1966.

———. *Colonialism 1870–1945: An Introduction*. London: Weidenfeld and Nicolson, 1981.

Franghia, G. "Report on Irrigation and Orange Growing at Jaffa," Great Britain: Foreign Office, *Miscellaneous Series no. 300, Reports on Subjects of General and Commercial Interest*. London: 1893.

Fruit Commission. *Report of the Fruit Commission*. Palestine: Government of Palestine, 1927.

Gal, Allon. "Brandies' View on the Upbuilding of Palestine." *Studies in Zionism* 6 (1982): 211–40.

Gaskin, Ian-William. *Economic Aspects of the Citrus Industry in Palestine 1918–1948*. Liverpool: Liverpool University, 1985.

Gera, Gershon. *The Unknown Generous Benefactor*. Tel Aviv: Modan, 1984 (Hebrew).

Gerschenkron, Alexander. *Economic Backwardness in Historical Perspective: A Book of Essays*. Cambridge: Harvard University Press, 1962.

Giladi, Dan. "The Private Farmers' Stand towards the Issue of 'Hebrew Work' in the Moshavot, 1904–1914." *Baderech* 6 (1971): 69–73 (Hebrew).

———. *Jewish Palestine during the Fourth Aliya Period (1924–29): Economic and Social Aspects*. Tel Aviv: Am Oved, 1973 (Hebrew).

———. "From Aliya to Aliya, between Growth to Crisis." In *The History of Eretz Israel: The British Mandate and the Jewish National Home*, ed. Yehoshua Porath and Ya'acov Shavit. Jerusalem: Keter, 1982, 161–73 (Hebrew).

———. "Industrial Enterprises in Eretz Israel in the 1920s." In *Jews in Economic Life*, ed. Nachum T. Gross. Jerusalem: Zalman Shazar Center, 1985, 347–55 (Hebrew).

Ginossar, Pinhas. "The Community of Hebrew Writers in Eretz-Israel, 1918–1923." *Iyunim Bitkumat Israel* 6 (1996): 479–534 (Hebrew).

Golan, Haim, ed. *The Poale-Zion Commission in Palestine—1920*. Ramat Efal: Yad Tabenkin, 1989 (Hebrew).

Goldwater, A. *Citrus Fruit in Palestine: Their Development and Prospects*. Jerusalem: Palestine Corporation, 1927.

Gonen, Amiram. "Urbanization of the 'Moshavoth' on Israel's Coastal Plain: Factors and Stages." In *Studies in Geography and History in Honour of Yehoshua Ben-Arieh*, ed. Yossi Ben-Arzi, Israel Bartal, and Elchanan Reiner. Jerusalem: Magnes Press and Israel Exploration Society, 1999, 461–77 (Hebrew).

Gorny, Yosef. "The Ideology of the 'Conquest of Labor.'" *Keshet* 10:2 (1968): 66–79 (Hebrew).

———. *Achdut Haavoda, 1919–1930: The Ideological Principles and the Political System*. Tel Aviv: Hakibbutz Hameuchad, 1973 (Hebrew).

———. *Zionism and the Arabs, 1882–1948: A Study of Ideology*. Oxford: Clarendon Press, 1987.

Gottschalk, Yehiel Alfred. *Ahad Ha-'Am and the National Spirit*. Jeruslaem: Hassifriya HaZiyonit, 1992 (Hebrew).

Graetz, Michael. *The Jews in Nineteenth-Century France: From the French Revolution to the Alliance israelite universelle*. Stanford: Stanford University Press, 1996.

Greenberg, Yitzhak. *From Workers' Society to Workers' Economy: Evolution of the Hevrat Ha'ovdim Idea 1920–1929*. Tel Aviv: Papyrus, 1987 (Hebrew).

Griliches, Zvi. "Hybrid Corn and the Economics of Innovation." In *The Economics of Technological Change: Selected Readings*, ed. Nathan Rosenberg. Great Britain: Penguin,1971, 211–28.

Gross, Nachum T. "Laying the Foundations." In *Banker for a Renewed Nation: The History of Bank Leumi Le'Israel*, ed. Nadav Halevi. Ramat Gan: Massada, 1977, 7–110 (Hebrew).

———. "Economic Changes in Eretz Israel at the End of the Ottoman Period," *Cathedra* 2 (1977): 111–25 (Hebrew).

———. "Private and Public Enterprise in Building Jewish Palestine According to Z. D. Levonteen." *Economic Quarterly* 30 (1983): 488–94 (Hebrew).

———. "The Economic Policy of the Mandatory Government in Palestine." *Research in Economic History* 9 (1984): 143–85.

———. "The Development of Agricultural Techniques in the Jewish Economy in Mandatory Palestine." In *The Development of Agricultural Technology in the 19th and 20th Centuries*, ed. Harald Winkel and Klaus Herrmann. St. Katarinen: Scripta Mercaturae, 1984, 157–68.

———. "Die Deutsche Palaestina-Bank 1897–1914." *Zeitschrift fuer Unternehmensgeschichte* 33:3 (1988): 149–77.

———. "Entrepreneurship of Religious and Ethnic Minorities." In *Juedische Unternehmer in Deutschland im 19. und 20. Jahrhundert*, ed. Werner E. Mosse and Hans Pohl. Stuttgart: F. Steiner, 1992, 11–23.

———. "Free Enterprise Versus Planned Markets," in Nachum T. Gross, *Not by Spirit Alone: Studies in the Economic History of Modern Palestine and Israel.* Jerusalem: Magnes Press and Yad Izhak Ben-Zvi, 1999, 355–65 (Hebrew).

———. "Citricultre until 1939: A Success Story?" In Nachum T. Gross, *Not by Spirit Alone: Studies in the Economic History of Modern Palestine and Israel.* Jerusalem: Magnes Press and Yad Izhak Ben-Zvi, 1999, 384–89 (Hebrew).

———. "The Absorption of University-Educated Immigrants in the 1920s and Hydrological Research." In Nachum T. Gross, *Not By Spirit Alone: Studies in the Economic History of Modern Palestine and Israel.* Jerusalem: Magnes Press and Yad Izhak Ben-Zvi, 1999, 273–86 (Hebrew).

———. "The Palestine Economy in the Final Decades of Ottoman Rule." In *Not by Spirit Alone: Studies in the Economic History of Modern Palestine and Israel.* Jerusalem: Magnes Press and Yad Izhak Ben-Zvi, 1999, 19–52 (Hebrew).

Gross, Nachum T., and Yitzhak Greenberg. *Bank Ha'poalim: The First 50 Years, 1921–1971*. Tel Aviv: Am Oved, 1994 (Hebrew).

Gross, Nachum T., and Jacob Metzer. "Public Finance in the Jewish Economy in Interwar Palestine." *Research in Economic History* 3 (1978): 87–159.

Gurevich, Dov, and Aharon Gerz. *Jewish Agricultural Settlement in Palestine*. Jerusalem: The Jewish Agency, Department of Statistics, 1938 (Hebrew).

Gvati, Chaim. *A Century of Settlement: The History of Jewish Agricultural Settlement in Eretz-Israel.* Tel Aviv: Hakibbutz Hameuchad, 1981 (Hebrew).

Halevi, Nadav. "The Economic Development of the Jewish Community in Palestine 1917–1947." Jerusalem: Maurice Falk Institute for Economic Research in Israel, Discussion Papers No. 7914, 1979 (Hebrew).

Hanotaiah. *Netanyah and What It Symbolizes*. Tel Aviv: Hanotaiah, 1932 (Hebrew).

———. *Orange Groves in Palestine*. Tel Aviv: Hanotaiah, 1933.

Hayami, Yujiro, and Vernon W. Ruttan. *Agricultural Development: An International Perspective*. Baltimore and London: Johns Hopkins University Press, 1971.

Hazen, William N. *The Citrus Industry of Palestine*. Washington, DC: United States Department of Agriculture, Bureau of Agricultural Economics, December 1938.

Hever, Hannan. *Poets and Bullies: The Evolution of the Hebrew Political Poetry in Palestine*. Jerusalem: The Bialik Institute, 1994 (Hebrew).

Herzl, Theodor. *The Jewish State: An Attempt at a Modern Solution of the Jewish Question*. London: Central Office of the Zionist Organization, 1936.

Hoofien, Eliezer Ziegfried, Zvi Viteles, and Mark Yaffe. *Recommendations for the Unification of the Jewish Citrus Industry's Cooperatives*. Tel Aviv: 1938 (Hebrew).

Horowitz, Dan, and Moshe Lissak. *Origins of the Israeli Polity: Palestine under the Mandate*. Chicago: University of Chicago Press, 1978.

Horowitz, David. The *Economy of Palestine and its Development*. Tel Aviv: The Bialik Institute, 1944 (Hebrew).

Hughes, Jonathan. "Public Sector Entrepreneurship." In *Favorites of Fortune: Technology, Growth, and Economic Development since the Industrial Revolution*, ed. Patrice Higonnet, David S. Landes, and Henry Rosovsky. Cambridge and London: Harvard University Press, 1991, 297–321.

Hurwitz, Dalia. "The Attitude of the Zionist Institutions and Leaders in Palestine to Jewish Labour during the Period of the Second Aliya (1904–1914)." *Zionism* 7 (1981): 95–134 (Hebrew).

Ilan, Zvi. "Holzmann's Partners in Establishing Kefar 'Etzyon in the 1930s." In *The Etzion Bloc from Its Inception to 1948 (Idan 7)*, ed. Mordechai Naor (Jerusalem: 1986): 63–74 (Hebrew).

Izakson, Eliahu. *A View from the Bridge Generation*. Tel Aviv: Eliahu Izakson, 1994 (Hebrew).

Kahn, Arcadius. "Notes on Jewish Entrepreneurship in Tsarist Russia." In *Entrepreneurship in Imperial Russia and the Soviet Union*, ed. Gregory Guroff and Fred Carstensen. Princeton: Princeton University Press, 1983, 104–24.

Kamen, Charles S. *Little Common Ground: Arab Agriculture and Jewish Settlement in Palestine, 1920–1948*. Pittsburgh: University of Pittsburgh Press, 1991.

Kaplansky, Shlomo. "Private Initiative and Rural Settlement." *Ha-ahdut* 5 (Summer 1914): 2–5 (Hebrew).

Kark, Ruth. "Transportation in Nineteenth-Century Palestine: Reintroducion of the Wheel." In *The Land that Became Israel: Studies in Historical Geography*, ed. Ruth Kark. New Haven, London and Jerusalem: Yale University Press and The Magnes Press, 1989, 57–76.

———. *Jaffa: A City in Evolution, 1799–1917*. Jerusalem: Yad Izhak Ben-Zvi, 1990.

———. "The Introduction of Modern Technology into the Holy Land (1800–1914)." In *The Archaeology of Society in the Holy Land*, ed. Thomas E. Levy. London: Leicester University Press, 1995, 524–41.

Karlinsky, Nahum. "California Dreaming: Adapting the 'California Model' to the Jewish Citrus Industry in Palestine, 1917–1939." *Israel Studies* 5:1 (2000): 24–40.

———. *Citrus Blossoms: Jewish Entrepreneurship in Palestine 1890–1939*. Jerusalem: Magnes Press, 2000 (Hebrew).

Katz, Shaul. *Sociological Aspects of the Growth and the Turnover of Agricultural Knowledge in Israel: The Emergence of Extra-Scientific Systems for the Production of Agricultural Knowledge 1880–1940*. Unpublished PhD dissertation, the Hebrew University of Jerusalem, Jerusalem, 1986 (Hebrew).

———. "Social Structure and Technological Innovation in Israeli Agriculture 1880–1930." *Scripta Hierosolymitana* 34 (1993): 48–70.

———. "On the Technological History of Palestine in the Late Ottoman Period: Three Case Studies." In *The Second Aliya: Studies*, ed. Israel Bartal. Jerusalem: Yad Izhak Ben-Zvi, 1997, 189–212 (Hebrew).

Katz, Yossi. *Ye Shall Grant a Redemption for the Land: The "Geula" Society for Land Acquisition 1902–1914*. Jeruslaem: Yad Izhak Ben-Zvi, 1987 (Hebrew).

———. "Private Zionist Initiative and the Settlement Enterprise in Eretz-Israel in the Early 1900's: 'Nationalistic Capitalism' of Private Capital." In *The Land that Became Israel; Studies in Historical Geography*, ed. Ruth Kark. New Haven, London and Jerusalem: Yale University Press and The Magnes Press, 1989, 275–86.

———. *Zionist Private Enterprise in the Building of Eretz-Israel during the Second Aliyah 1904–1914*. Ramat Gan: Bar-Ilan University Press, 1989 (Hebrew).

Katznelson, Berl. *Collected Writings*. Tel Aviv: Mapai, 1946–1950 (Hebrew).

Katzparovsky, Shimshon. "What One May Learn from California about Growing Citrus Trees in Palestine." *Ha-haklai* 5, 3 (Summer 1918): 7–12.

Kilby, Peter. "Hunting the Heffalump." In *Entrepreneurship and Economic Development*, ed. Peter Kilby. New York: The Free Press, 1971, 1–40.

Kimmerling, Baruch. *The Struggle for the Land: A Chapter in the Sociology of the Jewish-Arab Conflict*. Jerusalem: The Hebrew University of Jerusalem, 1974 (Hebrew).

———. *The Economic Interrelationship between the Arab and the Jewish Communities in Mandatory Palestine*. Cambridge: Center for International Studies, MIT, 1979.

———. *Zionism and Territory; The Socio-Territorial Dimension of Zionist Politics*. Berkeley: Institute of International Studies, University of California, 1983.

———. *Zionism and Economy*. Cambridge, MA: Schenkman, 1983.

Kislev, Yoav, and Willis Peterson. "Prices, Technology, and Farm Size." *Journal of Political Economy* 90 (1982): 578–95.

Kleinman, Rahel, Yehoshua Ben-Aryeh, and Dan Giladi. "The Beginnings of Netanya as an Agricultural Colony, 1928–1933." In *Netanya Book*, ed. Avshalom Shmueli and Moshe Brawer. Tel Aviv: Am Oved, 1982, 115–22 (Hebrew).

———. "Initial Development of an Urban Settlement alongside the Colony of Netanya, 1933–1939." In *Netanya Book*, ed. Avshalom Shmueli and Moshe Brawer. Tel Aviv: Am Oved, 1982, 123–38 (Hebrew).

Kolatt, Israel. *Ideology and the Impact of Realities upon the Jewish Labour Movement in Palestine, 1905–1919*. Unpublished PhD dissertation, The Hebrew University of Jerusalem, 1964 (Hebrew).

———. "Ha-po'el ha-Tsa'ir: From Conquest of Labor to Sanctification of Labor." *Ba-Derekh* 1 (1967): 29–61 (Hebrew).

———. *Fathers and Founders*. Tel Aviv: Institute for Contemporary Jewry, The Hebrew University of Jerusalem, and Hakibbutz Hameuchad, 1976 (Hebrew).

Kolatt, Israel et al. "Discussion: The 1929 Disturbances as a Turning Point." In *Jerusalem in Zionist Vision and Realization*, ed. Hagit Lavsky. Jerusalem: Zalman Shazar Center for Jewish History, The Center for the Study of Zionism and the Yishuv, Hebrew University of Jerusalem, 1989, 407–30 (Hebrew).

Kressel, Getzel. *Petah Tikva, Mother of Settlements, 1878–1953*. Tel Aviv: Municipality of Petah Tikva, 1953 (Hebrew).

Kuznets, Simon. *Modern Economic Growth: Rate, Structure, and Spread*. New Haven and London: Yale University Press; 1966.

Landes, David S. *The Unbound Prometheus: Technological Change and Industrial Development in Western Europe from 1750 to the Present*. Cambridge: Cambridge University Press, 1969.

———. "Introduction: On Technology and Growth." In *Favorites of Fortune: Technology, Growth, and Economic Development since the Industrial Revolution*, ed. Patrice Higonnet, David S. Landes, Henry Rosovsky. Cambridge and London: Harvard University Press, 1991, 1–29.

———. *The Wealth and Poverty of Nations: Why Some Are so Rich and Some so Poor*. London: Little, Brown, 1998.

Laskov, Shulamit. "The 'Moshavot' under the Tutelage of 'Hovve Zion' and Those Who Were Not." In *The History of the Jewish Community in Eretz-Israel since 1882: The Ottoman Period*, ed. Israel Kolatt, part one. Jerusalem: The Israel Academy for Sciences and Humanities, The Bialik Institute, 1989, 351–429 (Hebrew).

Lavsky, Hagit. *The Foundations of Zionist Financial Policy: The Zionist Commission 1918–1921*. Jerusalem: Yad Izhak Ben-Zvi, 1981 (Hebrew).

———. "Businessmen, Zionism, and the Building of the National Home: The Case of Salman Schocken." In *Jews in Economic Life*, ed. Nachum T. Gross. Jerusalem: Zalman Shazar Center, 1985, 357–72 (Hebrew).

———. "Has There Been an 'Essential' Connection between Jews and Capitalism?: The Case of Mandatory Palestine." In *Religion and Economy: Connections and Interactions*, ed. Menahem Ben-Sasson. Jerusalem: Zalman Shazar Center for Jewish History, 1995, 387–400 (Hebrew).

———. *Before Catastrophe: The Distinctive Path of German Zionism*. Detroit and Jerusalem: Wayne State University Press, The Magnes Press, Leo Baeck Institute, Jeruslaem, 1998.

———. "Theory and Praxis: The Agrarian Policy of the Jewish National Fund during the Mandatory Period." In *Studies in Geography and History in Honour of Yehoshua Ben-Arieh*, ed. Yossi Ben-Arzi, Israel Bartal, and Elchanan Reiner. Jerusalem: Magness Press and Israel Exploration Society, 1999, 438–60 (Hebrew).

Lazonick, William. "What Happened to the Theory of Economic Development?" In *Favorites of Fortune: Technology, Growth, and Economic Development since the Industrial Revolution*, ed. Patrice Higonnet, David S. Landes, Henry Rosovsky. Cambridge and London: Harvard University Press, 1991, 142–58.

Leroy-Beaulieu, Paul. *De la Colonisation chez les peuples modernes*. Paris: Guillaumin, 1902.

Levontin, Zalman David. *To the Land of Our Fathers*. 3 vols. Tel Aviv: Eitan and Shohsany, 1924–1928 (Hebrew).

Liberecht, 'Akiva. "Agricultural History of Petah Tikva." *Bustenai* 10:27 (1938): 35–38 (Hebrew).

Lissak, Moshe. "'Critical' Sociology and 'Establishment' Sociology in the Israeli Academic Community: Ideological Struggles or Academic Discourse?" *Israel Studies* 1:1 (1996): 247–49.

Lockman, Zachary. *Comrades and Enemies: Arab and Jewish Workers in Palestine, 1906–1948*. Berkeley, Los Angeles, and London: University of California Press, 1996.

Loew, Yehuda. *Agricultural Economy*. Tel Aviv: Am Oved, 1957 (Hebrew).

Mansfield, Edwin. "Technical Change and the Rate of Imitation." In *The Economics of Technological Change: Selected Readings*, ed. Nathan Rosenberg. Great Britain: Penguin,1971, 284–315.

Mashke, Shulamit. *The Problem of Marketing the Palestinian Oranges in Europe*. Jerusalem: 1937.

Mendelsohn, Chaim. *International Trade in Oranges: Competition for Export Markets*. Unpublished PhD dissertation, University of California, Berkeley, 1958.

Metzer, Jacob. "Rational Management, Modern Business Practices, and Economies of Scale in the Ante-Bellum Southern Plantation." *Exploration in Economic History* 12 (1975): 123–50.

———. *National Capital for a National Home 1919–1921*. Jerusalem: Yad Izhak Ben-Zvi, 1979 (Hebrew).

———. "Technology, Labor, and Growth in a Dual Economy's Traditional Sector: Mandatory Palestine, 1921–1936." In *Technical Change, Employment, and Investment*, ed. Lennart Joerberg and Nathan Rosenberg. Lund: Lund University, 1982, 159–70.

———. "What Kind of Growth? A Comparative Look at the Arab Economies in Mandatory Palestine and in the Administered Territories." *Economic Development and Cultural Change* 40 (1992): 843–65.

———. *The Divided Economy of Mandatory Palestine*. Cambridge: Cambridge University Press, 1998.

Metzer, Jacob, and Oded Kaplan. *The Jewish and Arab Economies in Mandatory Palestine: Product, Employment and Growth*. Jerusalem: Maurice Falk Institute for Economic Research in Israel, 1990 (Hebrew).

Michaely, Michael. *Foreign Trade and Capital Imports in Israel*. Tel Aviv: Am Oved, 1963 (Hebrew).

Mokyr, Joel. *The Lever of Riches: Technological Creativity and Economic Progress*. New York and Oxford: Oxford University Press, 1990.

———. "Technological Inertia in Economic History." *Journal of Economic History* 52 (1992): 325–38.

———. *The Gifts of Athena: Historical Origins of the Knowledge Economy*. Princeton: Princeton University Press, 2002.

Nathan, Robert R., Oscar Gass, and Daniel Creamer. *Palestine: Problem and Promise—An Economic Study*. Washington DC: Public Affairs Press, 1946.

Nawratzki, Curt. *Die Juedische Kolonisation Palaestinas*. Muenchen: Reinhardt, 1911.

Near, Henry. "Ideology, Anti-Ideology, and Historiography." *Cathedra* 18 (1981): 124–29 (Hebrew).

Ofer, Gur. *The Service Industries in a Developing Economy: Israel as a Case Study*. New York; F. A. Praeger, 1967.

Olson, Mancur. *The Logic of Collective Action: Public Goods and the Theory of Groups*. New York: Schocken, 1965.

Oppenheim, I. D. "The Development of Cultural Practices in the Citrus Industry." *Hadar* 11:1 (1938): 40–42 (Hebrew).

Oren, Amiram. *Agricultural Experiments and Experiences in the Zionist Settlement, 1882–1914*. Jerusalem: Yad Izhak Ben-Zvi and the Volcani Center, 1993 (Hebrew).

Owen, Roger. "Economic Development in Mandatory Palestine: 1918–1948." In *The Palestinian Economy: Studies in Development under Prolonged Occupation*, ed. George T. Abed. London and New York: Routledge, 1988, 13–35.

Pappé, Ilan. "Zionism as Colonialism: A Comparative Look at Mixed Colonialism in Asia and America." In *From Vision to Revision: A Hundred Years of Historiography of Zionism*, ed. Yechiam Weitz. Jerusalem: Zalman Shazar Center, Herzl Institute for the Study of Zionism, Haifa University, and the Cherrick Center for the Study of Zionism, the Yishuv and the State of Israel, the Hebrew University of Jerusalem, 1998, 345–65 (Hebrew).

Pardess Society. *Annual Reports of the Pardess Society* (various years), also listed under Pardess Cooperative Society of Orange Growers (est. 1900), Ltd., Jaffa, Report. Reports appear under this title until 1939/40, inclusive (Hebrew).

Pardess Society—Statutes. *Statutes of the Pardess Society in Petah Tikva, Colonies of Judea, and Samaria* (before 1914) and various statutes of the society under various names (up to 1939).

Penslar, Derek J. *Zionism and Technocracy: The Engineering of Jewish Settlement in Palestine, 1870–1918*. Bloomington and Indianapolis: Indiana University Press; 1991.

———. "Zionism, Colonialism, and Postcolonialism." *Journal of Israeli History* 20:2–3 (2001): 84–98.

Plessner, Yakir. *The Political Economy of Israel: From Ideology to Stagnation*. Albany: State University of New York Press, 1994.

Poshter, A. *The Development of the Citrus Marketing Cooperative Societies in the 1937/38 Season*. The Anglo-Palestine Bank, March 1939.

Ram, Uri. "Society and Social Science: Establishment Sociology and Critical Sociology in Israel." In *Israeli Society: Critical Perspectives*, ed. Uri Ram. Tel Aviv: Breirot, 1994, 7–39 (Hebrew).

———. "Historiographical Foundations of the Historical Strife in Israel." *Journal of Israeli History* 20:2–3 (2001): 43–61.

Rav-Krakotzkin, Amnon. "History Textbooks and the Limits of Israeli Consciousness." *Journal of Israeli History* 20:2–3 (2001): 155–71.

Redlich, Fritz. "Entrepreneurship in the Initial Stages of Industrialization (with special reference to Germany)." In *Entrepreneurship and Economic Growth*. Cambridge: The Harvard University Research Center in Entrepreneurial History, 1954, Section B.

———. "Innovation in Business: A Systematic Presentation." In Fritz Redlich, *Steeped in Two Cultures: A Selection of Essays*. New York and Evanston: Harper and Row, 1971, 163–71.

———. "Entrepreneurial Typology." In *Steeped in Two Cultures: A Selection of Essays*, ed. Fritz Redlich. New York and Evanston: Harper and Row; 1971, 112–32.

Registrar of Cooperative Societies in Palestine. *Report by the Registrar of Cooperative Societies in Palestine on the Development during the Years 1921–1937*. Jerusalem: Government Printing Press, 1938.

Reichman, Shalom. "The Evolution of Land Transportation in Palestine, 1920–1947." *Jerusalem Studies in Geography* 2 (1971): 55–90.

Reichman, Shalom, and Shlomo Hasson. "A Cross-Cultural Diffusion of Colonization: From Posen to Palestine." *Annals of the Association of American Geographers* 74: 1 (March 1984): 57–70.

Reisen, Esther. "High Hopes: The Beginning of Moshe Smilansky's Public Career." Unpublished MA thesis, Tel Aviv, Tel Aviv University, 1982 (Hebrew).

Reuveny, Jacob. *The Administration of Palestine under the British Mandate, 1920–1948: An Institutional Analysis*. Ramat Gan: Bar-Ilan Univesity Press, 1993 (Hebrew).

Rogers, Everett M. *Diffusion of Innovations*. New York: Free Press, 1962.

Rokach, Isaac. *Tales of the Jaffa Orange Groves*. Ramat Gan: Massada, 1970 (Hebrew).

Rosenberg, Nathan. *Technology and American Economic Growth*. New York: Harper and Row, 1972.

———. *Inside the Black Box: Technology and Economics*. Cambridge: Cambridge University Press; 1982.

———. *Exploring the Black Box: Technology, Economics, and History*. Cambridge: Cambridge University Press; 1994.

Rosenberg, Nathan, and L. E. Birdzell. *How the West Grew Rich: The Economic Transformation of the Industrial World*. New York: Basic Books, 1986.

Ruppin, Arthur. *Agricultural Settlement of the Zionist Organization in Palestine (1918–1924)*. Tel Aviv: Dvir, 1925 (Hebrew).

———. *Thirty Years of Building in Palestine*. Jerusalem: Schocken, 1937 (Hebrew).

Ruttan, Vernon. "Usher and Schumpeter on Invention, Innovation, and Technological Change." In *The Economics of Technological Change*, ed. Nathan Rosenberg. Great Britain: Penguin, 1971, 73–85.

———. *Technology, Growth, and Development: An Induced Innovation Perspective*. New York and Oxford: Oxford University Press, 2001.

Ryerson, Knowles A. "The Horticultural Possibilities as Especially Related to Agricultural Colonization." *Reports of the Experts: Submitted to the Joint Palestine Survey Commission*. Boston: Daniels Printing Co., 1928, 237–397.

———. *The World is My Campus*. University of California, Davis, Shields Library: The Oral History Center, 1977.

Salmon, Yoseph. "The Urban Ashkenazi Community in Palestine, 1880–1903." In *The History of the Jewish Community in Eretz-Israel since 1881: The Ottoman Period*, ed., Israel Kolatt, part one. Jerusalem: The Israel Academy for Sciences and Humanities, The Bialik Institute, 1989, 539–619 (Hebrew).

Sawer, E. R. *A Review of the Agricultural Situation in Palestine*. Palestine: Government of Palestine. Department of Agriculture & Fisheries, 1922.

Schama, Simon. *Two Rothschilds and the land of Israel*. London: Collins, 1978.

Schultz, Theodore W. *Transforming Traditional Agriculture*. New Haven and London: Yale University Press, 1969.

Schumpeter, Joseph A. "Economic Theory and Entrepreneurial History." In *Change and the Entrepreneur: Postulates and Patterns for Entrepreneurial History*. Prepared by the Research Center in Entrepreneurial History, Harvard University. Cambridge: Harvard University Press, 1949, 63–84.

———. "The Instability of Capitalism." In *The Economics of Technological Change: Selected Readings*, ed. Nathan Rosenberg. Great Britain: Penguin, 1971, 13–42.

———. "The Fundamental Phenomenon of Economic Development." In *Entrepreneurship and Economic Development*, ed. Peter Kilby. New York: The Free Press, 1971, 43–70.

Sela, Avrahahm. "The Disturbances at the Wailing Wall (1929): A Turning Point in Jewish-Arab Relations?" In *Jerusalem in Zionist Vision and Realization*, ed. Hagit Lavsky. Jerusalem: Zalman Shazar Center for Jewish History, The Center for the Study of Zionism and the Yishuv, Hebrew University of Jerusalem, 1989, 261–78 (Hebrew).

Shafir, Gershon. *Land, Labor, and the Origins of the Israeli-Palestinian Conflict, 1882–1914*. Berkeley and London: University of California Press, 1996.

Shalev, Michael. *Labour and the Political Economy in Israel*. Oxford: Oxford University Press, 1992.

———. "Time for Theory: Critical Notes on Lissak and Sternhell." *Israel Studies* 1:2 (1996): 170–88.

Shapira, Anita. *Futile Struggle: The Jewish Labor Controversy, 1929–1939*. Tel Aviv: Hakibbutz Hameuchad, 1977 (Hebrew).

———. *Berl Katznelson: A Biography*. Tel Aviv: Am Oved, 1983 (Hebrew).

———. *Visions in Conflict*. Tel Aviv: Am Oved, 1989 (Hebrew).

———. "The Politics of the Jewish Community in Palestine, 1918–1939." In *The History of the Jewish Community in Eretz-Israel since 1881: The Period of the British Mandate, Part Two*, ed. Moshe Lissak, Anita Shapira, and Gavriel Cohen. Jerusalem: The Israel Academy for Sciences and Humanities, The Bialik Institute, 1994, 1–70 (Hebrew).

———. "Politics and Collective Memory: The Debate over the 'New Historians' in Israel." *History & Memory* 7:1 (1995): 9–40.

———. "Sternhell's Complaint: 'Nation Building or a New Society' by Zeev Strenhell." *Yiunim Bitkumat Israel* 6 (1996): 553–67 (Hebrew).

———. "The Strategies of Historical Revisionism." *Journal of Israeli History* 20:2–3 (2001): 62–76.

Shapiro, Yonathan. *The Historical Ahdut ha-'Avoda: The Power of Political Organization.* Tel Aviv: Am Oved, 1975 (Hebrew).

Shavit, Yaacov. "The Jewish Community in Eretz Israel in the Mandatory Period: Between a Society and Polity." *Cathedra* 14 (1980): 7–16 (Hebrew).

———. *The Textiles in Eretz Israel 1854–1956: From a Pioneering to a Leading Industry.* Tel Aviv: Israel Textile Association, 1992 (Hebrew).

Shavit, Yaacov, and Dan Giladi. "The Role of the Dairy Farm in the Development of Jewish Settlement during the Mandatory Period." *Cathedra* 18 (January 1981): 178–92 (Hebrew).

Sherman, Neal, and Daniella Heller. "The Establishment and Institution of the Farmers' Federation of Eretz Israel." *Cathedra* 58 (1990): 155–71 (Hebrew).

Shilo, Margalit. "Dr. Arthur Ruppin and Settlement in the Mountain Area: His Attitude toward the 'To the Mountains' Affair and the First Kefar Etzion." In *The Etzion Bloc from Its Inception to 1948 (Idan 7)*, ed. Mordechai Naor. Jerusalem: 1986, 41–52 (Hebrew).

———. *Experiments in Settlement.* Jerusalem: Yad Izhak Ben-Zvi, 1988 (Hebrew).

Shilony Zvi. "German Antecedents of Rural Settlement in Palestine up to World War I." In *The Land that Became Israel; Studies in Historical Geography*, ed. Ruth Kark. New Haven, London, and Jerusalem: Yale University Press and The Magnes Press, 1989, 196–214.

Shimoni, Yaacov. *The Arabs of Palestine.* Tel Aviv: Am Oved, 1947 (Hebrew).

Slushtz, Nachum. *Dr. Aharon Meir Mazie: His Life and Activities.* Jerusalem: Jewish Exploration Society of Israel, 1935 (Hebrew).

Slutsky, Yehuda. *The History of the "Hagana" Organization*, vol. 2. Tel Aviv: Hassifriya Haziyonit and Ma'arachot, 1965 (Hebrew).

Smilansky, David. *A City Is Born: The Story of Tel Aviv and Eretz-Israel in the Second Aliya Period, According to the Letters of David Smilansky as Published in the Press in Russia.* Ed. and Introduction Yossi Katz. Tel Aviv: Ministry of Defense Publishing House, 1981 (Hebrew).

Smilansky, Moshe. "On Questions of Colonization in Palestine." *Hashiloah* 14 (1904): 291–301, 432–42 (Hebrew).

———. "With the Flow of Phrase." *Ha-Herut* 5:293–97 (1913), in segments (Hebrew).

———. "Jewish Colonization and the Fellah." *Palestine and Near East Economic Magazine* 5 (1930).

———. *Writings of Moshe Smilansky*, 12 volumes. Tel Aviv: Farmers' Federation, 1934–1937 (Hebrew).

———. *In the Fields of Ukraine: A Story in Five Parts.* Tel Aviv: Massada, 1944 (Hebrew).

———. *In the Wilderness: A Story*. Tel Aviv: Massada, 1947 (?) (Hebrew).

———. *Rehovot: Its Sixty Years of Life, 1890–1950*. Rehovot: Rehovot Local Council, 1950 (Hebrew).

———. *In the Shadow of the Orchards: A Story*. Tel Aviv: Massada, 1951 (?) (Hebrew).

———. *Resurrection and Holocaust: A Story*. Tel Aviv: Massada, 1953 (Hebrew).

———. *Among the Vineyards of Judea: A Story*. Tel Aviv: Massada, 1954 (Hebrew).

———. *Birth Pangs: A Story*. Tel Aviv: Massada, 1954 (Hebrew).

———. "Personal Questionnaire." *Genazim* 7 (April 1979): 276–79 (Hebrew).

Smith, Barbara J. *The Roots of Separatism in Palestine: British Economic Policy, 1920–1929*. Syracuse: Syracuse University Press, 1993.

Stein, Kenneth W. *The Land Question in Palestine 1917–1939*. Chapel Hill and London: University of North Carolina Press, 1984.

Sternhell, Zeev. *Nation-Building or a New Society?: The Zionist Labor Movement (1904–1940) and the Origins of Israel*. Tel Aviv: Am Oved, 1995 (Hebrew).

Stoll, Steven. *The Fruits of Natural Advantage: Making the Industrial Countryside in California*. Berkeley, Los Angeles, London: University of California Press, 1998.

Sussman, Zvi. *Wage Differentials and the Equality within the Histadrut: The Impact of Egalitarian Ideology and Arab Labor on Jewish Wages in Palestine*. Ramat Gan: Massada, 1974 (Hebrew).

Tarrant, John R. *Agricultural Geography*. New York: John Wiley and Sons, 1974.

Temin, Peter. "Entrepreneurs and Managers." In *Favorites of Fortune: Technology, Growth, and Economic Development since the Industrial Revolution*, ed. Patrice Higonnet, David S. Landes, and Henry Rosovsky. Cambridge and London: Harvard University Press, 1991, 142–58.

Teveth, Shabtai. *David's Jealousy: The Life of David Ben Gurion*, vol. 2. Jerusalem and Tel Aviv: Schoken, 1980 (Hebrew).

Thalmann, Naftali. "Introducing Modern Agriculture into Nineteenth-Century Palestine: The German Templers." In *The Land that Became Israel: Studies in Historical Geography*, ed. Ruth Kark. New Haven, London, and Jerusalem: Yale University Press and The Magnes Press, 1989, 90–104.

———. *The Character and Development of the Farm Economy in the Templar Colonies in Palestine, 1869–1939*. Unpublished PhD dissertation, Hebrew University of Jerusalem, Jeruslaem, 1991 (Hebrew).

Thirtle, Colin G., and Vernon W. Ruttan. *The Role of Demand and Supply in the Generation and Diffusion of Technical Change*. Chur: Hardwood Academic Publishers, 1987.

Tidhar, David. *Encyclopaedia of the Pioneers and Builders of the Yishuv*. Tel Aviv, various years.

Tocqueville, Alexis de. *Democracy in America*. Translated by George Lawrence. New York: Harper and Row, 1966.

Todaro, Michael. *Economic Development*. New York and London: Longman, 1994.

Tolkowsky, Shmuel. *Citrus Fruits: Their Origin and History throughout the World*. Jerusalem: The Bialik Institute, 1966 (Hebrew).

Troen, Ilan S. "American Experts in the Design of Zionist Society: The Reports of Elwood Mead and Robert Nathan." In *Envisioning Israel: The Changing Ideas and Images of North American Jews*, ed. Allon Gal. Jerusalem and Detroit: The Magnes Press and Wayne State University Press, 1996, 193–218.

Tzahor, Zeev. "Farmers and Labourers of the Second Aliya in Petah Tivka." *Cathedra* 10 (1979): 142–50 (Hebrew).

Ulitzur, Abraham. *National Capital and the Building of the Country, 1918–1937*. Jerusalem: Keren HaYesod, 1939 (Hebrew).

United States Department of Agriculture, Bulletin no. 1234. *Organization and Development of a Cooperative Citrus-Fruit Marketing Agency*. Washington, DC, 1924.

Vashitz, Joseph. *The Arabs in Eretz Israel*. Merhavia: Sifriat Ha-Poalim, 1947 (Hebrew).

Viteles, Harry. "Expansion of the Orange Industry in Palestine." *Bulletin of the Palestine Economic Society* 3:1 (1928): 3–104.

———. "Survey of the Citrus Industry in Palestine." *Hadar* 11:1, 2–3 (1938): 1–43.

Von Thünen, Johann Heinrich. *The Isolated State*. Trans. Carla M. Wartenberg. Ed. Peter Hall. Oxford: Pergamon, 1966.

Waldinger, Roger. "The Two Sides of Ethnic Entrepreneurship." *International Migration Review* 7 (1993): 692–701.

Waldinger, Roger, Howard Aldrich, Robin Ward et al. *Ethnic Entrepreneurs: Immigrant Business in Industrial Societies*. Newbury Park, CA: SAGE Publications, 1990.

Weinberg, Isaac. "The Orange Trade of Jaffa." *Palestine Economic Society* 3 (1924): 3–32.

Weitz, Raanan. *From Peasant to Farmer: A Revolutionary Strategy for Development*. New York and London: Columbia University Press, 1971.

Wilkansky, Yizhak. "On the Expectations for Capital." *Hapoel Hatzair* 6:35 (1913): 3–6.

———. *Hapoel Hatzair* 6:36 (1913): 5–9 (Hebrew).

Wilken, Paul H. *Entrepreneurship: A Comparative and Historical Study*. Norwood, NJ: Ablex, 1979.

Wolochonsky, D. "Rishon Lezion in the Lights of the Statistics." *Mis'har Ve-Ta'asiyah* 8:4–5 (1930): 39–62 (Hebrew).

Yakhin Cooperative Society. *Statutes*. Tel Aviv: Yakhin, 1929 (Hebrew).

———. *Four Years of Activities of the Yakhin Cooperative Society for Agricultural Cultivation, 1927–1930*. Tel Aviv: Yakhin, 1931 (Hebrew).

———. *Annual Reports*, various years.

Yedidyah, Shmuel. *Citrus Growing*. Tel Aviv: Hasadeh, 1937 (Hebrew).

Yizhar, S. *Mikdamot*. Tel Aviv: Zmora-Bitan, 1992 (Hebrew).

Yudilowitz, David, ed. *Rishon Lezion, 1882–1941*. Rishon Lezion: Carmel Mizrahi, 1941 (Hebrew).

Zagorodsky, Melekh. *The Citrus Book: Manual for Orange Growers, Workers, Experts, and Nurseries Owners*. Tel Aviv: Hahaklai, 1929 (Hebrew).

Zakif, Shmuel. *Diaspora and Homeland*. Magdiel: Magdiel Municipality, 1954 (Hebrew).

Zipperstein, Steven Jeffrey. *Elusive Prophet: Ahad Ha'am and Origins of Zionism*. Berkeley: University California Press, 1993.

Index

Aaronsohn, Aaron, and Selig Soskin, *Die Orangengärten von Jaffa*, 89–91, 93–98, 100, 164
Ab'd al-Rachman, Selim, 161
Abu Kishk, 159–160
Acre, 81–82
'Agr, 158
Agricultural Experiment Station (Mikve Israel), 148, 204
Agricultural Experiment Station (Rehovot), 62, 147–148
Agricultural Experiment Stations (governmental), 183
Ahad Ha'am (Asher Hirsch Ginsberg), 24, 26, 31–37, 52
Anglo Palestine Bank (APB), 25, 60, 64, 213–215
Anglo Palestine Company (APC). *See* Anglo Palestine Bank (APB)
Apfelbaum, Meir, 51, 57, 60, 102–108, 204
Arab (Palestinian) citrus industry, 8–9, 49–50, 80–82, 89–102, 123, 130–131, 148–152, 155–164, 191–198, 220; exporters, 199, 201–203; geographical distribution of, 17, 80–82; profitability of, 136, 148–152, 167–173; technological innovations in, 89–98, 123, 130–131, 134–140, 148–152, 155–164
Assis, 1, 141
Associated Orange Growers, 192–193
bahar, 94, 100, 103, 116
Bayt Dajan, 161
Beidas, Sa'id, 203
Ben-Ami, Oved, 76–78, 218
Benei Binyamin, 38, 76
Ben-Gurion, David, 30, 37, 39–41
berara, 69–70, 140–142
beyara, 92–93, 191
al-Bitar, Abdel Rauf, 203
"Boazes," 31–37, 41–43
Bodenheimer, Simon, 145, 148
Brazil (citrus industry), 174
Brill, Abraham, 51, 143, 188
British Mandate for Palestine, 9, 36–37
Bushnaq, Otman, 161
Bustan, 209
by-products industry, 140–142

California Fruit Growers Exchange, 187–188
California model, 111–112, 117–123, 127, 131–134, 187–190, 206, 218
Citrus Control Board, 15, 216
Citrus Fruit Committee, 182–183, 201–202
Citrus Marketing Board, 15, 203, 216
Clark Powell, H., 119–120, 188
colonialism and Zionism, 7–8, 25–45
colonialism (European), 7–8, 26–29, 34–36
"conquest of labor," 1–4, 7, 35, 40–41. *See also* "Hebrew" labor

cooperatives, 193–198, 204–216. *See also under* marketing
cultivation, 94–95, 100, 103–105, 116–122

decoville, 122, 138, 140
Degania, 36, 56, 72
"dispossessed" Arabs, 157
Divrei Mordechai, 89

'Ein Ganim, 36
electric system, 125–126
Emek Hefer. *See* Hefer Valley
entrepreneurship: and Jewish citrus growers, 2, 57–83 *passim*, 217–220; "Schumpeterian," 10–11, 58–65, 78, 218; theories of, 9–12, 45, 218–219

Farmers' Federation, 16, 23–24, 79, 114, 203–204, 209
Fellmann, Aharon Leib, 50, 89–95, 98
Fourth Aliya, 6, 37–41, 67–68
Franghia, George (report of), 89–91, 95–96, 98

Gan Chaim, 159
Gardinger, Moshe, 159
Gaza, 81–82
German-Templer, 96, 101
Ginsberg, Asher Hirsch. *See* Ahad Ha'am
government (of Palestine), 9, 146–148, 181–186, 189, 197, 215–216, 219; advertising regulations, 186 ; inspection regulations, 183–186
grapefruit, 142–144
groundwater (availability of), 50, 54, 91–94, 98–99, 123–130

Hadar, 205–206
Hadera, 24
hadovile, 138
Haifa (port of), 69–70, 199
haluts (*halutsim*), 1, 22, 42–43
Hanotea, 76–78, 118, 218
Hazen, William, 133, 136, 142
"Hebrew" labor, 1–4, 7, 22–23, 30–31, 37, 39, 42–44, 113–115, 218. *See also* conquest of labor
Hefer Valley, 118
Hibbat Tsiyyon (and Hovevei Tsiyyon), 23–26, 37, 57
Holzmann, Shmuel Zvi, 62–63
Hoofien, Eliezer, 213
Horin, Yehuda, 113, 218

irrigation and irrigation systems, 92–94, 98–103, 122, 130–134
Italy (citrus industry), 174

Jacobsohn (family), 63–64
Jacobsohn, Eliezer, 63
Jacobsohn, Zalman, 63–64, 192–193, 201, 212
Jaffa, 2, 49, 57, 80–82; port of, 69
Jaffa Citrus Exchange, 62, 64, 198–203
Jaffa Orange Syndicate, 207–216
Jaffora, 141
Jalant (brothers), 203
Jewish Colonization Association (ICA), 36, 41, 63
Jewish (Zionist) citrus industry, 12–16, 167–179; attempts of unification, 189–216; geographical distribution of, 17, 49–83; ideological traits, 21–45, 112–115; profitability of, 106–110, 113–115, 119–122, 136, 148–154, 167–173, 189–190; technological innovations in, 98–106, 112–154
Jewish labor. *See* "Hebrew" labor
Jewish National Fund (JNF), 5–6, 28, 34–35, 37, 158
Judea (region), 18, 52–56, 66–67, 116–117, 125, 131, 197

Kafr 'Ana, 161
Kafr Saba, 158–159
Katznelson, Berl, 3–4, 37–40
Kfar Saba, 4, 66, 158

Labor Movement (Zionist): agricultural settlements of, 5–6, 37–40, 72; attitude towards private enterprise, 9,

21–23, 37–41; economic activities and institutions of, 37–40, 72, 204, 212–216 *passim*, 218; ideology of, 2–5, 9, 21–23, 30–31, 33; 37–41
Land Settlement Ordinance (1928), 158–160
Levontin, Zalman David, 28
locomobile, 100, 103
Lubman Haviv, Ya'akov, 157–160, 162

Machnes, Gad, 22, 76–78
al-Maghar, 158
Mandatory government. *See* government (of Palestine)
manège, 100, 103
marketing methods, 191–204; on the tree, 191–192, 207; commission, 192–193; cooperative, 193–198, 207–216
markets, 169, 173–179; British, 169, 173–177, 193–198 ; Central Europe, 169, 174, 178–179, 194; German, 169, 178, 194; Western Europe, 173, 178, 194
Mashke, Shulamit, 173–174
al-Masri, Taher, 161
Mazie, Aharon Meir, 58–60
Merkaz, 62, 194–195
Miller, Tovia, 60–61, 114, 116, 158
Miska, 159
moshava, *moshavot*, 1–4, 6, 18, 30–31, 34–36, 65–72, 158–164
musha', 157–159

Nablus, 161
al-Na'ni, 158
National Farmers' League, 114
national settlement (Zionist), 21–45 *passim*
Nawratzki, Curt, 52–54, 101–102
Nes Tsiyyona, 53, 158, 214
Netanya, 76–78, 117, 161
nuriyya, 92–93, 98–99

Oppenheim. Y. D., 117

Palestine Corporation, 213
Palestine Economic Corporation, 212
Palestine Jewish Colonization Association (PICA), 5, 41
Palestinian-Arab citrus industry. *See* Arab (Palestinian) citrus industry
Pardess, 52, 57–58, 62, 168–173, 188, 191, 193–216 *passim*
Pardess Syndicate, 209–216
Pascal, Peretz, 51, 57, 60, 106, 116
Petah Tikva, 3, 36, 43, 51–60, 67, 71, 125, 128–129, 161, 193–194, 208
picking and packing, 135–142
planting distances, 94, 104–105, 119–120, 134
Pri-Taz, 214
private settlement (Zionist), 21–45 *passim*
profitability. *See under* Arab (Palestinian) citrus industry; Jewish (Zionist) citrus industry
propagation, 94–95, 105–106, 134–135
pumping system, 92–94, 98–103, 122–130

Qalqilya, 161
al-Qubayba, 158, 161

Rabinovitz, Lazar, 192–193
al-Rahim, Muhammad 'Abd, 203
railroad system, 68–71
Ram, Baruch, 76–78
Rantiya, 161
Rehovot, 24, 53, 56, 60–61, 67–68, 158, 161–164
Rishon Lezion, 24, 53–56, 67, 96, 161
Rok, Alfred, 203
Rokach (family), 57–58, 204, 208, 218
Rokach, Isaac, 97, 192, 199, 201, 204–216, 218
Rokach, Shimon, 57–58, 64, 204, 218
Rothschild, Baron Edmund de (administration of), 5, 24, 35–36, 41, 51–56, 58, 60, 99, 101
Ruppin, Arthur, 30–31, 36, 39, 56
Ryerson, Knowles, 116–117, 119–121, 127, 190

Sachs, Ephraim, 60, 65
Salama, 161

Samaria (region), 18, 52–56, 66–67, 78–79, 116–119, 125–130 *passim*, 198
Saqiya, 161
Sarafand al-Kharab, 158, 161
Sawer, E. R., 189
Shak'a, Ahmed, 161
Sharon (region), 18, 66–67, 75–79, 116–119, 125–131 *passim*, 197
al-Shaykh Muwanis, 160, 203
Shneir (family), 192–193
Smilansky (family), 23–25
Smilansky, Moshe, 22–45 *passim*, 53–56, 63–64, 118, 122, 158, 202–216 *passim*, 218; attitude towards the Palestinian-Arabs, 26–27, 42–44, 160–161; on Zionist colonization, 25–45 *passim*; on the role of private entrepreneurship, 26–37, 41–45, 114–115
Smilansky, Ze'ev, 24, 161–164
South Africa (citrus industry), 174–178
spacing. *See* planting distances
Spain (citrus industry), 174–178
species and varieties, 16–17, 142–144

al-Taji, 'Abd al-Rahman, 202–203
al-Taji, Shukri, 203
technological innovations, 87–91; adoption and diffusion, 88–89, 98–106, 127–128, 133; in Arab citriculture. *See under* Arab (Palestinian) citrus industry, in Jewish citriculture. *See under* Jewish (Zionist) citrus industry
Tel Aviv, 30, 34–35, 43, 58–59, 71–72, 158–160
Tel Mond, 117
Templars. *See* German-Templer
Thünen, Johann Heinrich von, 50–53, 70
Tnuva Export, 40, 75, 213–214
Tocqueville, Alexis de, 21–22
Tolkowsky, Shmuel, 22, 49, 61–62, 64, 201, 208
transport system, 68–71
Tulkarm, 77, 80–82, 161

Umm-Khalid, 161
Union, 194–195
United States (citrus industry), 174
USDA Economic Bureau Report. *See* Hazen, William

Viteles, Harry (Zvi), 188–189, 212–213

Weizmann, Chaim, 41–43, 62, 159
Wells. *See* pumping system
Wilkansky, Yitzhak, 33

Yaffe, Mark, 213
Yakhin, 40, 72–75, 117–119, 127, 131–132, 213–214, 218
Yedidya, Shmuel, 123
Yehudiya, 161
Yishuv, 1–2, 44; demography of, 6; occupational structure of, 4–6

Zagorodsky, Melekh, 116–117
Zarnuqa, 158, 161–164
Zikhron Ya'akov, 53, 66, 96
Zionist Organization, 5–6, 38, 116
Zu'aiter, A'adel, 161